THE OCEAN UNIVERSE

THE OCEAN UNIVERSE

Robert L. Straitt

Edited by

Nadine L. Straitt

Dedication

I would like to dedicate this book to two people who over the many years of research and writing have inspired the development of the concepts described herein and who have supported my efforts through their sacrifices to make this work possible.

First my loving wife Nadine, who without her continued encouragement as well as giving up of many-many hours of her own and our families time to support my efforts, this work would not of been possible.

Second to my father, Lloyd L. Straitt, who from my earliest childhood recollections to his passing in in 1993, encouraged and challenged me to look beyond the surface of what we call reality and ask the questions why and how. It was my father, who first introduced me to the mechanics and effects of gyroscopic precession, which first sparked my interest in the study of Relativity from my early high school days. Although he was there to help in my early stages of writing this book, it is unfortunate he did not see it completed.

Table of Contents

Preface

March 2, 1993. As we approach the dawning of a new era we can feel the excitement of the unexpected begin to overwhelm us. What can we expect from technology in this new era? What can we expect from the anticipated "New World Order"? We should be concerned with, "What Should We Expect from this new Era?" We are about to begin more than a new millennia, we are at the threshold, a beginning, of a new level of technology and science, that not only is far advanced of our greatest achievements to date but transcends our very comprehension of the nature and the mechanics of the universe.

To crossover this threshold and begin a new journey requires us to forgo the fundamental values we have adopted and basic thought patterns that have served us so well and brought us steadfastly to the end of this plateau. Much as stone-age man had to give up his stone-hammer and cave for iron and a change in the necessitated social and economic systems. We must be willing to forgo the technology we have grown accustom to and the stagnation it now imposes. In order to step across the growing chasm we, as individuals, must be able to assess our science as an integral part of the fabric of the universe and recognize events not for their impact today but for their consequences on tomorrow.

A little over thirty years ago, this book was first put to ink, which literally means hand written with pen and ink, as computer technology was only in its infancy at the time. During this phase of my career I was working with some of the most advanced technology development programs that the government had, and the computers we had on our desktops would probably not operate the key board on my current laptop today. Looking back on the last thirty years and the developments in science and technology that have taken place, I have to wonder if perhaps the information and theories proposed herein were before their time. Would they have been ignored? I think so because we had just begun to cross the threshold into the digital information age, and revisiting the old methods and thoughts just was not vogue in scientific circles of that time.

In 2012, the movie Battleship was released on the world. A motion picture describing the attack on earth by alien beings, who were made aware of our planets existence, by our own attempts to try to contact alien life on other planets in our galaxy similar to our own. Though the special effects were great and the story line a little too familiar, what came to my mind as I watched, was the failure to be able to defend our world, even though our military had all the latest technologies in modern science available to them. In the end, it was the old and now antiquated science and technology, which was brought to bear on the enemy to defeat the aliens. Sometimes Hollywood has a way of accentuating reality in ways that were never intended. Could this scenario be analogous to the current state of our modern physics?

When I started out on this topic some thirty years ago now, modern physics was already old; Einstein had released his theory on General Relativity in 1916 imposing a new order on science for well over 60 years at that point. Soon, it will be one hundred years since Einstein changed the dynamics of our perspectives on science, and yet the scientific community has still not been able to embrace one of the most fundamental perspectives of his work. This book will lay out a number of reasons that challenges us to reconsider and adopt that perspective.

I hope you enjoy reading this book, I have left much of the older material as it was originally written, I found it interesting to notice the differences in literary styles between then and now. I have also strived to write in a style that will be comfortable to the widest audience base. Thus, the hard academic may be bored with its simplicity and lack of complex mathematics and the novice might be overwhelmed with the abstract style of natural philosophy. Enjoy!

Introduction

This book is about understanding our universe; what makes it work, how it affects our existence, and how we can affect ourselves by efficient application of these natural universal rules. In order for us to do this, we must have a thorough knowledge of the very fabric of the universe and the laws that have established the order.

When we wake up in the morning, we experience a very simple and regular routine in all appearances, the sun rises in the east, and later in the day we see it set in the west. At least most people would argue that this is a fair assessment of the observed phenomena, although I have met some that will not accept this description. One could say that this was a static event, in that it continues without detectable change. However, two other possibilities exist. One being that the routine is breaking down, becoming more chaotic with time. The second is that the routine began as a random event and is continuing to stabilize with time. Three very plausible and distinct possibilities for any given instant with any one of them being true, however, how will we know which one is in fact true?

To understand the unknown one must have a fixed point of reference from which to build ones knowledge. Much as with writing a computer program certain basic rules apply in all situations and the rules are not allowed to be broken or the whole logic of the system would fall apart. When a computation must be made that does not fit any of the existing rules, as they stand, a combining of rules, in a specific sequence must be accomplished in order to compute the desired function.

If for example, we were to look at a functional computer program as a finished product. The most obvious thing we would see is apparent violations of rules, known to be absolute. For example suppose we had an IF statement. A simple IF statement is a function in which the computer would determine if a certain predefined condition exist or does not exist. A simple yes or no! Should the condition exist one action should be taken and if the condition does not exist, another action should be taken. No other possibilities are allowed. The only allowed possibilities would look like this:

If situation is true take action x.

If situation is not true, take action y.

A computer program that can select between thousands or millions of choices and make a selection would then appear to violate the rule that allows for only one choice and two possible actions.

As an observer to this, we can make one of three cases.

1) The rule was wrong or misstated.

2) The rule allows for special cases.

3) We just do not know how the rule has been used in this instance to arrive at these results.

As a human observer, we most likely would shy away from observation 3 as it highlights our own lack of knowledge. Something about our humanity causes us to gravitate away from a position in which we have to admit a lack of knowledge, perhaps it is just an innate fear of the unknown, associated with fight or flight responses. More often than not, we would most likely try to rationalize the observation with a new rule or even worse a "special case" which makes an exception to an existing rule under certain circumstances. The ability of inanimate matter to ascertain that certain complex conditions exist for it, relative to an observer is one thing that has always amazed me. I question the ability of the inanimate to recognize that it (the inanimate matter) must react to a specific event in a specific way to conform to the criteria of a special case, established because of human observation.

The most compelling reason of concern with the use of special cases in modern physics is the implied inference that the inanimate objects involved in the action have a sense of being or intelligence. While not necessarily like a human, rather more like a simple computer. Take for instance the relationship between two beams of light. With our current understanding of physics, we have a paradox. The relative speed between the beams must not exceed the speed of light and the observed speed of the two beams by observers at any point must be the same. Thus, each beam must exhibit a speed of C and a speed of C/x, to multiple observers, at multiple positions, at the same time. Thus to explain this paradox in modern physics special cases, such as time dilation are employed.

If we were to use the same example, only substituting bullets for light beams, the speed of the bullets would be observed to remain constant and the relative speed of the bullets to each other would reflect their actual relative speeds. In physics, because it is a form of mathematics, all events are subject to fixed rules (laws). Except those events whose observed actions are contrary to these rules, in which case, the rules are suspended so as to explain the observed inconsistency, thus maintaining the apparent legitimacy of the system of mathematics being employed, rather than explain the event.

Whenever a law of nature needs to be regularly exempted, in order for it to explain apparent observations; history shows that more often than not the law was, either originally misstated or it has been incorrectly applied in the subject incident. To fix many of these cases, the laws are simply restated or the rules and ways under which they are applied are modified so that all included observations then become consistent with the law as it is accepted.

An often-overlooked approach to explaining phenomena is to reanalyze the event being observed by breaking it down into its smallest freestanding and complete components. Then by evaluating each component attribute of the event against existing laws to verify their compliance. If this is accomplished and each component of the event complies with an existing law, then the apparent paradox is just that, apparent. If any component of the event does not obey existing laws then that component should be reevaluated to ensure that it is the smallest possible component of the event. If the event component cannot be broken down into subcomponents and the actions of the component are accurately recorded, then the law in question is evaluated to verify that it is stated accurately. If necessary, the law can then be rewritten to include this new variation, as well as, other known events in a consistent manner.

Just as a computer applies fixed rules in a complex but systematic chain of discrete addition functions at the machine language level, resulting in what appears to be an impossible multiplication, at the more abstract high programming language level. Although many of us may be familiar with such procedures as, Karatsuba multiplication, Toom–Cook multiplication, Fourier transformation methods, and Gauss's complex multiplication algorithm, utilized in higher order programming languages. We know that most computers perform multiplications functions in base 256 procedures, yet ultimately the most primitive add-shift functions of the binary machine language are still depended upon, underneath each of these more abstract processes, to complete the multiplication tasks. Thus a simple set of the laws of physics may be applied to explain complex, apparently paradoxical events.

For centuries, man has held the same basic thought patterns and these patterns have served him well, having brought him to the end of an era. From the discovery of fire to that of atomic fusion, every man has perceived the world and studied it from the same perspective. We can no longer continue with this outlived approach. The rate of technological advance is doubling, even as you read this book. It is now more important than ever to reorganize the way we perceive knowledge and how we apply that knowledge or we will reach a "dead end". What type of dead end? A technological dead end, in which we are unable to advance beyond a certain level of technology, versus the more simplistic refinement of technology we currently use. We are at the very fringes of where our current technology base can take us. Much as

Stone Age man was limited by his technology base of stone and animal skins, so Information Age man can become limited by his tool box of intellectual rules and ideas, if doesn't ascend beyond his current understanding of the universe in which he lives.

This is not to say that technology has reached its climax however. Rather, it is to emphasize that further significant development in technology will only follow a drastic restructuring of our understanding of science. The bottom line is there is only so much you can do with a stone knife! To further illustrate this imagine a medieval alchemist being brought into a modern chemical research facility or imagine a prehistoric cave man who doesn't understand fire witnessing an atomic bomb denote. In their respective worlds they could survive, learn, and prosper but in our world, the science is beyond their grasp. Much as the taming of fire ushered in a new era of technology and understanding, the close of that era, (which we still live in) will usher in a new era that will be as wonderful and complex in its own right.

One may argue that we live in an era that is several eras removed from the cave man. Refinements in technology however, do not equal establishment of a completely new thought processes. Obviously, we take for granted many of today's conveniences that our old friend the cave man could never have even imagined. However, our every endeavor is governed by the same motivator as was his. We still go out to gather food and we still go out to gather fuel for our fires. We get up in the morning and prepare our meals over an open flame, go to work, come home, prepare another meal over an open fire. Our homes are still heated with chemical combustion. The material our homes are made from and the way we construct them today, is basically the same as it has been for thousands of years. Understanding that our current construction may appear more elegant, we must remember that planed lumber and 2x4s versus roughhewed logs is a technology refinement issue not a change in the scientific thought process.

Today as we ride to work in our luxury automobile, we are riding on the very same technology the Pharaoh used in ancient Egypt, the wheel. When we look at all forms of material handling from goods to humans, we find that the wheel or some refined form of it still plays the leading role. Today we are only beginning to reach beyond the barrier or our understanding of the universe and dabble in new technologies such as, magnetic levitation. However, we do not understand the basic science of the technology and even our understanding of its characteristics is extremely limited.

A wonderful example of our reaching the limits of new science and thought is the restrictions and licensing we have put on our ideas. Today our institutions of higher reasoning are much more concerned with plagiarism, cheating, profitable copyrights, and layout formatting according to one publishing association's format or another. Use of so-called peer reviewed materials at the academic level is enforced to protect profitable intellectual standings by eliminating and censoring controversial thoughts that may call into question or replace existing intellectual property rights owned and controlled by academic elitists. We no longer are training the majority of our students to think; rather we are training them to memorize the works of their professors without question and then consistently cite that work in numerous academic papers, journals, and books to increase the esteem of the professor, rather than the knowledge of the student (Cline, 2012).

It was only in the 18th century, with the development of large commercial printers and publishers that concepts of copyrights and intellectual property began to be converged with plagiarism. And it was only it was then that a master's inspiration was slowly converted from knowledge for social growth, into property of intellectual monopolies that guarantee wealth to the elite, while restricting the inspiration and creativity of reuse by individual thinkers that the copyright laws were intended to protect (Lynch, 2006). Contrary to what some scholars, such as Lise Buranen and Alice Myers Roy contend in their book *Perspectives on Plagiarism* and *Intellectual Property in a Postmodern World*, plagiarism was not historically always in disdain as Buranen and Roy suggest (Buranen & Roy, 1999). Historically plagiarism was a form of honorarium and to see ones thoughts, styles, and ideas in the work of one's students was the greatest honor and satisfaction one could hope for. Before the advent of the copyright laws designed to protect authors from the monopolistic book printers, literary works were generally considered to be in what the philosopher Horace described as *"publica materies"* or "in the public property" (Will, 1963).

"Harding's best points first, he reminds us that Quintilian was the chief bearer, to the Renaissance, of a conviction which had been so central in antiquity that it was seldom explicit. In literate, traditional societies-like those of Greece from Homer through the fifth-century, and of pre-Christian Rome –literature is public property, what Horace called 'pulica materies'. As we know, there were in antiquity no copyright laws; the greatest danger in literary copying was bad taste." (Will, 1963)

Just as fire and the wheel changed ancient civilizations, so must our concepts of not only science and nature change but also our expectations of civilization must change to allow for the introduction of a new technology into our lives. We need to redefine our expectations and our boundaries to create a framework in which change and advancement is the desired norm in our lives, in our government, and most importantly in our economic structure. For us to enjoy the benefits of new technologies as they become available we must have a business environment that is structured to profit from change, rather than from consistency. This requires a mindset of the public that is focused on the benefits that technology can bring and a social system that is flexible to variations. If we are approaching the limits of our current science and knowledge base, then we will soon begin to see an increase in possessiveness towards the current knowledge and a reluctance on the established scientific community to allow for new thought that may result in a new science and social structure to transcend the current body of knowledge.

The Ocean Universe

Chapter 1: The Development of Modern Science

Claudius Ptolemaeus (Ptolemy) was a Greek astronomer of the Second Century CE who wrote a treatise called "Almagest" which explained the observed movements of the planets through the sky from the perspective of an earth centered observer. Over time this concept became known as the as the Ptolemaic system. Ptolemaic created a mechanical model of the observed universe, accurately replicating the observed movements of the sun, planets, and stars. The accuracy of the Ptolemaic model was such that it was accepted by the church, as the actual structure of the universe, until the 16th century, when the heliocentric theory (sun centered) was gradually accepted. This model was based on the principle that each of the planets moved in epicycles (tiny circles) while they orbited the earth. The accuracy of the Ptolemaic model was such that it was still in use in some schools during the late Nineteenth Century to predict astronomical events.

Ptolemy had started out to explain the motions of the universe as observed. Studying the apparent motions of the heavenly bodies is somewhat of a lesson in Relativity, as well as astronomy, because of the lack of a common reference point. Ptolemy had to overcome one significant paradox in his observations, the apparent doubling back of the planets. This doubling back is called retrograde motion and is actually caused by the Earth's rotation, as the planets do not stop their forward motion. Ptolemy felt that this retrograde motion would be the significant factor in determining which theory of the universe (Sun centered or Earth Centered) was correct and set out to isolate an answer. Ptolemy carefully reviewed known data of planetary motion and then went on to conduct careful observations himself. As a means of replicating these observed astronomical events, he developed his mechanical model that mimicked the observed movements of the planets, stars, and sun.

Ptolemy however, was caught up in religious and political controversy of his day. It was under the influence of these religious and political powers that Ptolemy's observational model was used to validate and expand upon previous earth-centered theories of the universe in support of non-scientific goals. As a result, a model that Ptolemy intended to be used as a tool of scientific study was turned into a proof of a controversial theory. We might remember that some 300 year earlier Aristarchus proposed a sun centered theory of the universe, at theory that had been widely accepted in the Greek world. Hipparchus in the first century BCE, countered the sun centered model by proposing an earth-centered solution that would predict the movements of the planets, in apparent movement around the earth, as observed from a point on the earth. It was with Hipparchus' theory that the religious and political leaders of Ptolemy's time promoted their beliefs and with Ptolemy's model was able to establish the earth centered theory so firmly.

For over fourteen centuries, scientific developments were directed by the popularized and institutionalized misbelief that the earth was the center of the universe. During this time, many great discoveries were made and man achieved significant technological advancement. Unfortunately, even major errors in science are not readily identifiable and often because of non-scientific reasons, considerable effort will be expended to retain a scientific theory that has entrenched its self in the minds of man. It has been said that empires have been held at stake by the discoveries of science. Even after Copernicus published his treatise "De Revolutionibus Orbium Caelestium" in 1543, which mathematically proved that the sun was in fact the center of the solar system the powers to-be resisted the change that they, rightly, felt would eventually lead to their downfall.

In the world of science, new theories regularly replace existing beliefs only to be replaced by newer theories. Each transition is marked by a significant upset in the scientific community. What is even more interesting is that regularly the new theory turns out to be not as comprehensive as originally believed. In these cases, the truth turns out to be a hybrid product of previous and current theories that is more spectacular then was ever anticipated.

So how is a new theory developed? Most major developments in science come as response to a need. These needs have not always been as technologically orientated as what we are accustomed to today. In fact, a traditional motivator was merely the inquisitive human mind. Often religious or political goals presented the impetus for scientific endeavors, and science itself can often be self-perpetuating. When one reviews a string of scientific thought that lead to a theory being established, one must take into account both the sociological and technological environment in which the theorist conducted his work. This procedure often is not practiced however, and more often than not, new theories are proposed in such a manner as to discredit or completely do away with previously accepted materials.

Many questions confront humankind and the answers to these questions usually raise new and more perplexing questions. In order for one to rationally attempt to find meaningful answers, one must attempt to isolate a workable subset of related questions that can be addressed by a singular hypothesis. This is not an easy task. Albert Einstein once stated, "The theorist's method involves his using as his foundation general postulates or "principles" from which he can deduce conclusions. His work thus falls into two parts. He must first discover his principles and then draw the conclusions which follow from them." (Einstein, 1954)

A theorist works in quite a different world than that of other scientists and engineers. A theorist must explain the unexplainable within the parameters of the known. This is quite different from applying the known to new applications. To arrive at a theory then, one must first have a goal to be reached. The explanation of an observed event is such a goal. An observed event can also serve as the foundation from which to develop "principles" on which to base one's investigations. It is important to remember that the principles a theorist uses to lay a foundation of a postulate are not always a valid part of the end theory. If we recall Ptolemy's mechanical model, we remember that it was constructed as a tool to develop a postulate and was never really intended to be the end theory itself. As theorists or as enthusiastic students we should remember that the final expression of the truth through a theory is what matters, not the initial accuracy of the postulates.

As you proceed through this, book the realization that we are today working with theories of physics that have not yet reached their end states, will become more self-evident. As such, the defense of the current language (postulate) of these theories should take a back seat, to the advancement of the overall theory's achievement of expressing a final truth. Every postulate is subject to constant re-investigation and no part is sacred until the fabric of the whole has been woven and tried. The real joy of theoretical physics is that we are allowed to wander a little from side to side, as we try to discover and follow a defined path.

Chapter 2: Relativity, Einstein and Modern Man

In this chapter, we will look at, without fully restating the principles of, both Special and General Relativity. Then we will explore where Einstein was headed, where he had been, and how he may have gotten there if he was to start today. Modern man is a relative term in itself; however, it is used here to define today's contemporary physicist, a group who hope they understand Einstein but who often are only modern day Ptolemaists.

Einstein, much like Ptolemy, had to face a world that seemed to be in contradiction with his science. Having studied the known facts of his day, Einstein was amazed at how well organized the universe was for such a large and complex system. Einstein required a model to help him bring the immensity of the universe into his laboratory. Unlike Ptolemy however, Einstein used a mathematical model rather than a mechanical one. To Einstein's disappointment, others propelled his model forward as the final solution, the long awaited answer. To his dismay, they pronounced Quantum Mechanics the logical successor to Relativity. When in reality a ball falling from a train was but a single step in an infinitely longer journey to find the answers he sought.

From the simpler problem of a ball dropped from the window of a moving train to that of a passenger moving around within the train, to the nature of the constant speed of light, Special Relativity defines motions of objects relative to each other. Like Ptolemy's Almagest, Special Relativity is based on observable phenomena and although takes into account time, it retains its roots in a three dimensional system of logic. We really need not to update relativity but to update the ways in which we use both relativity and Newtonian Mechanics to assist us in understanding a four dimensional nature of our universe, which includes time within its fundamental understanding.

All of us who have had elementary physics should be familiar with Einstein's railroad car. The concepts of relativity are neatly described by using a moving railroad car and the embankment as two unique frames of reference from which to view an event. The apparent observations used in this hypothetical event have been the bases for the well-established concepts (laws) for the behavior of light, the dilation of time, and most significantly the interrelationship of matter and energy.

According to Einstein using the defined models of Newtonian Mechanics, we would plot the path of the ball from either the train or the embankment depending on our point of reference. We would not be able to correlate, directly, the recorded path of the ball as seen by one observer against the path of the ball as seen by the observer from the other frame of reference. The problem, for Einstein was how to relate common observations between observers with different observational perspectives. Einstein has answered that problem to a degree with Relativity. As we know, Einstein's approach called for the ball to have two apparent paths depending on the frame-of-reference of the observer and the relationship between those paths is defined by the relationships between the two frames of reference. This supposed capability of crossing from one frame-of-reference to another appeared, at the time of its being defined, to be a significant improvement over Newtonian Mechanics.

Using a four-dimensional rather than a three dimensional perspective we would first define a beginning and ending point for our calculations. In other words, when time is included in our analytical model, we must specify when our calculation begins and when it ends to ensure an accurate definition of events. If time is left open, we could be confronted with an event that happens much after our observer leaves the area (i.e. a little boy finds the ball a day later and carries it home).

For our analysis, the time-frame being considered will begin when the ball is released and end when it strikes the ground. We have now defined the two most important events within our example. The ball is released at point/time 'A' (once freed, becomes an object independent of the railroad car) and the ball strikes the ground at point/time 'B'. Although we have now coordinates for these points, we do not have insight into the travels of the ball. Now, if we are trying to predict the travel of the ball, from our pre-established observation point on the embankment, we can assume that whatever path the ball takes or appears to take, that path will be influenced by the velocity and position of ball with respect to the ground at the time of release. Another possible condition is that we are analyzing a prior event, one that has already occurred with respect to our place in space/time. In this case, the ball release point, the point at which it struck the ground, and the ball's path of travel are already defined by the event and can be reviewed by us.

If we were to provide a means of determining position along the railroad bed, such as fixed markers, we could then correlate the observations by all our observers in both time and space. These delimiters will serve to mark for us a particular area in space and time. All defined events will happen within these two markers. The crossing of the markers by the ball will be determined by each observer independently and then correlated for comparison of observations. In essence, what we are doing by this is merging separate frames of reference into a common frame of reference. It is very analogous to merging separate cells on a spreadsheet, as not only the information content but the separately configured envelope that contain it are merged into one cell with consistent configuration rules for all data contained within the merged cell.

Returning to our original problem, let us look at that railroad car moving along a straight section of tracks at a constant velocity. We will have a person hold a ball out one of the windows in such a manner as to be able to see the ball, the ground, and the railroad bed, what would that person observe? What would a person on the embankment observe? What would a person on the railroad bed observe?

The person within the railroad car holding the ball would see the following:

1) The ball apparently stationary relative to the side of the railroad car;

2) The ball apparently moving at a constant velocity relative to the railroad bed;

3) The ball apparently moving at a constant velocity to the embankment.

These observations can be supported by Newtonian Mechanics, the Distributed Law of Equivalency, and our own sense of logic. If A is rigidly attached to B and B is moving with a Velocity V then A is moving at a Velocity V also.

If A=B and B=V then A=V.

The observer on the embankment would observe:

1) The ball appearing stationary, relative to the side of the railroad car;

2) The ball apparently moving at a constant velocity to the railroad bed;

3) And, the ball apparently moving at a changing velocity, relative to the observer, as the railroad car approaches and then moves away.

If the observer on the embankment has an optical scanner capable of observing the ball only in a direction perpendicular to the railroad bed, but the scanner range transcended the distance the ball traveled along the railroad bed, then observations would be of a ball stationary to the railroad car moving at a constant velocity.

The observer on the railroad bed would observe:

1) The ball apparently stationary to the side of the railroad car;

2) The ball apparently moving at a constant velocity to the embankment;

3) The ball moving at an apparently changing velocity as the railroad car approaches and then moves away from the observer on the railroad bed.

Now if our person in the railroad car were to drop the ball at the instant the ball is over the preselected marker point what would our observers see? The observer on the embankment would see the ball begin to fall along a path described by a parabolic curve. The observer on the railroad bed would have an unusual view of the whole event and would see the ball descending perpendicular to the railroad bed while moving along the bed at a decreasing velocity. Both of these observations are consistent with the original observations provided by Einstein. Now let us look at what the observer in the railroad car would see. As the ball leaves our holders hand, it begins to descend but also it begins to decelerate and drop back from its original point relative to the railroad car. Thus, the ball as observed by the person in the railroad car also follows a path described by a parabolic curve. At this point, we find a difference in our observations from those described by Einstein.

Einstein described the observations of the person in the railroad car as the ball apparently falling in a path perpendicular to the holder's hand. "I stand at the window of a railway carriage which is travelling uniformly, and drop a ball on the embankment, without throwing it. Then, disregarding the influence of the air resistance, I see the ball descend in a straight line." (Einstein, 1961) The ball would follow such a path only if its forward acceleration relative to down ward acceleration was to remain constant. One can verify this observation by holding a heavy object (one with a low surface area to negate air resistance) below your eye, as you look down and while moving your head in a constant velocity drop the object. You will notice that the object does not stay directly under your eye but tends to drop back as it falls, while your head moves forward.

This is a very important concept because it shows that regardless of our observers' frames of reference, they all observe the ball follow the same trajectory, albeit each observation was from a different perspective. Mathematically the events could be rotated and transposed and the curve of decent would be the same for all observers. In the early 1980's it would be harder to demonstrate this to be true but today with our advances in 3D modeling and animation software it would be an easy to create a 3D representation of a ball falling from a railroad car. Gravity will create "drag" on the object as soon as it is released. Freed from the acceleration of your hand the object now maintains its own reference frame that is independent from your hand. A ball dropped from a train will experience the same transition to a new reference frame and will be subject to the gravitational effects of the earth. Such gravitational forces are perpendicular to the earth's center and follow the laws of General Relativity as well.

Why does not Einstein reach the same conclusion? First Einstein was limited by his technology and secondly Einstein was breaking new ground. We must remember that Einstein's ultimate objective was to unify all of the known forces of his time and Relativity was just Einstein's first step in that direction. Therefore, we must, as Einstein did, allow for the possibility that the original postulates laid forth in Relativity may not be outcome intended. Had Einstein had the benefit of a Doppler radar to make his observations with, he would have been presented with much different observational data. Rather than throw away his postulates though, we can use them as the shell from which to begin reconstruction of our physics. Einstein was trying to explain a completely new approach to viewing the universe, but he was explaining that new vision from a platform firmly attached to the contemporary physics of the day. His vision was that of a universe that existed in a physical sense but also in sequential sense. This sequential ordering of events is what the fourth dimensional aspect of the universe is all about. Today we can look back at Einstein's work, apply new understandings of time to his original postulates, and gain a more in depth understanding of the events he was describing. It is often said that "Hindsight is 20/20", and here we find ourselves with the unique advantage of hindsight and to some extent, along with the obligation to use that advantage to continue the journey that Einstein started.

What we can learn from this new look at the railroad car hypothesis is that events are relative to the frame-of-reference of the observer, as Einstein stated, and they are also absolute. What this infers is that our ball must have an absolute path through the universe that begins at a specific point in space-time and ends at specific point in space-time. Suppose we had to place a fixed object along the railroad bed so that the ball when dropped from the railroad car would hit the object. If we were required to deal with multiple paths of the ball as observed from various positions on the train or on the ground it would be next to impossible to establish a scenario in which the ball could hit our target because of the infinite number of possible observations and combinations of observations. In practice however we know that we can hit our target with the ball and at least one application of the methodology for doing so has been proven by military bombers and today's hypersonic airborne weapon systems.

This is not an effort to dismiss Relativity's concept of different frames of reference, rather it is an attempt to relate the concept of multiple frames of reference with the Newtonian Model of absolute space. We live and operate in a frame-of-reference in which all the rules of Newtonian Mechanics operate perfectly, yet the principles of relativity are valid for certain cases within our frame of reference. Because of this apparent overlap, it would appear logical that both sets of physics may be partial solutions. If so then what we need to do is unify these two theories in a comprehensive manner. Another words by maintaining Newton's Laws and appending the apparent observational effects described by Relativity one can begin to formulate a hypothesis that does not require special cases but explains events in a straight forward and linearly predictable manner. This is consistent with the general ideas set forth by Einstein in his later years, as he tried to piece together a Grand Unification Theory. A very plausible argument may then follow that in order for Newtonian Mechanics and Relativity to coexist the universe must have a form of an absolute frame of reference. A reference point from which all other frames of reference are in motion. We will explore this previously ignored concept from a new perspective and look a critical element that both Relativity and Newtonian Mechanics have in common. This concept of a master reference frame has generally gone unrecognized, although in many aspects it is at the very heart of not only Relativity and Newtonian Mechanics as well as Quantum- Physics.

The Ocean Universe

Chapter 3: Relativity Versus Reality

Before going into the mechanics of how Relativity and Newtonian physics can be harmoniously blended to develop a unification theory we should reexamine some of the accepted conclusions, paradoxes and proposed proofs of relativity. We will analyze these and present some interesting arguments about some of the most widely accepted paradoxes. I think it is important to revisit momentarily, a very basic assumption made earlier. Inanimate objects (clocks, rockets, stones, light, etc.) cannot ascertain the environment surrounding them. Inanimate objects by definition do not have the capability for that type of intelligence.

It has probably become somewhat apparent that I have left Quantum Theory out of the discussions for the most part. This is because much of contemporary Quantum Mechanics is extrapolated from, if not based on, principles described by relativity. Relativity however, does not impose Quantum Mechanics, it is simply that Relativity, as left by Einstein, is easily described in such a manner as to make it seem a logical predecessor to Quantum Mechanics. As we are interested in a more preliminary discussion, concerned with the most basic issues of relativity, we would only confuse these discussions by including Quantum Theory at this point.

Relativity proposes that events happen in frames of reference that are often independent of each other. For example in an elevator, with no windows, moving at a constant velocity, one would be unable to tell if the elevator was in motion or not. If only a single porthole was provided, with a view of the inside of the elevator shaft only, then the person inside would be unable to ascertain if the elevator was moving or if the elevator shaft was moving. From the perspective of relativity, it makes little difference, which is moving the elevator or the building. From the perspective of the person who has to step out the elevator door, it could make a significant difference. This is the concept of reality! Forty floors down to the street is a long way to fall.

If events, such as motion, happen relative to each other, as describe by relativity and to some extent Newtonian Mechanics, then those events must be relative to each other in totality. If two events have any additional relativity to them then they must both be relative to at least one additional frame of reference. The concept of isolated relativity has a very narrow range of applications in the real universe, as most events can be described relative to two or more frames of reference. Let us look at a traditional example (paradox) that is used to explain/dispute the principles Special Relativity.

If a spaceship were to take off from the earth and travel to a distant star at some relativistic speed and then return to the earth at a relativistic speed (relative to the earth), it is proposed that an astronaut within the spaceship would have aged, at a rate less than that of an earth observer (The Twins Paradox). This on face value is contrary to Relativity, which states that the event viewed from the earth and from the spaceship should be isometric. That is if only the earth and the spaceship existed, and then neither the astronaut nor the observer on the earth would be able to distinguish who is moving or who is stationary. The argument goes something like this. One twin (the brother) rides a rocket from earth to some distant star and back at a relativistic speed. The other twin (the sister) stays on earth. When the rocket arrives, back at earth the twin on the rocket has aged considerably less than the twin that remained on earth. The paradox here being that Relativity states that the movement of the rocket relative to the earth is reversible and therefore one could say that the earth moved relative to the rocket and the twin on the earth should therefore be younger. The answer to the paradox is explained in this manner. Because the rocket changes course at some point in its trip and heads back toward the earth then dilation of the clock will be on the rocket only. Thus, acceleration is an element of time dilation that results in the aging of the traveling twin as predicted.

In Euclidean geometry, the shortest total distance between two points is achieved by the traveler that does not change direction. All indirect paths are longer than this minimum. In space-time, the longest total interval occurs for the traveler that does not change direction. For all travelers who change direction, the total interval is shorter than this maximum.

Before we accept this explanation on face value, let us break the event down into all its component parts. As we said earlier sometimes, what appears to be an impossible event is actually a combination of several sub events that by themselves obey all the rules. If only two frames of reference existed, that of the clock on the earth and that of the clock in the rocket we could then address this problem in its most basic form. We will call the earth object A and we will call the rocket object B. By using Newtonian Mechanics, we could conclude that if B applied a force on A, so as to propel away from A, then the force applied on B by A must be equal to the force applied by B. [A = B so then to get B with a Velocity V respective to A then B+f/2=A-v/2 or B+f=A] Notice that in effect the Velocity of B is achieved by the exertion of half the total required force by B and the other half by A. As Newton's Third Law of Motion states, "for every action there is an equal and opposite reaction". Notice the similarity to Newton's Third Law of Motion and Special Relativity, The world of Special Relativity is much simpler that the gravity ridden world of General Relativity, yet it's more confusing to conceptualize. Everything is foot-loose and there are not definite points of reference like the gravitational centers of stars and planets. Restated this means that two events of motion isolated within their own frames of reference occurs relative to each other. Another way to state this is that, object A can be viewed as being in motion from object B and object B can be viewed as being in motion, from object A.

Without a fixed frame-of-reference relative to A and B it is impossible for an observer on A or B to tell which object is moving, which is standing still, or if they are both moving. One could then argue that for our twins, each of them would be moving relative to the other and the effects of slowing clocks would be equalized. They would both be the same age when they meet again. This is where the concept of changing direction (acceleration) comes into the equation. We thus have to as then if it is this really a valid concept?

Suppose our objects A and B are two identical, space ships rather than a spaceship and the earth. This will remove any question of one object greater mass being the determining factor in deciding, which object is traveling and which is staying behind. Following the logic of the twin paradox one of the space ships will travel out some considerable distance turn around and return to the other. Upon returning to the original point the twin on-board, the traveling spaceship will not have aged as much as the one on the space ship that stayed behind. I therefore place the following argument. "If the space ships have no frame-of-reference other than themselves, then the twins will be the same age when they meet again." How is this possible and still be consistent with the concepts outlined in Relativity?

First without a third frame-of-reference it is impossible to tell what the actual movement of either ship is. Only the relative movement between the ships can be concluded. This being the case, it can be argued that either ship A or ship B was in motion or that both ship A and ship B were in motion. This is not a scenario that one could expect to find, as it would eliminate the very space in which, ships A and B are supposed to be traveling in. However, it does show by extremes that the basic principle of Relativity, the relationship between two frames of reference, is valid. In this scenario each clock, as observed by a viewer in the other frame of reference, would appear to be going slower than a clock in that viewer's frame of reference. The net effect being that the twins would age equally. Because this is purely a hypothetical situation, this apparent canceling out is correct. If a different conclusion could be reached, from this hypothetical model, then logic of relativity would be flawed. However as we will see later, the existence of this frame of reference, isolated from other frames may not be possible in our universe, except under very extreme conditions.

A second and more interesting argument is based on not only Special Relativity but General Relativity as well. Suppose our spaceship has a special propulsion system that enables it to travel at very close to the speed of light C minus some infinitely small number). The maximum speed of the ship is such that time, for a traveler aboard the ship, would all but stand still. A second-of-time on board the ship may equal a billion or more years on earth. With this capability (and enough fuel), the ship could travel an almost infinite distance without the traveler aging appreciably. Now according to the accepted solution to the Twins Paradox the traveler is identified from the stay-behind by the fact that the travelers' ship changes course and returns to the point of origin. Before we go any further, let us explore this concept a little more.

The "Twins Paradox", as we remember is as follows. A ship leaves earth at a relativistic speed, on board is one of two twins (the brother), the other twin (the sister) stays on earth. When the ship returns to earth at some later date, the twin on-board the spaceship has not aged as much as the twin left on the earth. The basis for this assumption is, as the spaceship approached and maintained relativistic speeds, time within the spaceship frame-of-reference was proceeding at an appreciably slower rate, then that of time for the twin on earth. The argument is that according to relativity, each set of events is relative to the other. In other words the earth bound twin would perceive the spaceship riding twin's clock would be slower and the spaceship-riding twin would perceive the earth bound twin's clock to be running slower. The counter argument is that only the traveler's clock would be affected and the traveler can be distinguished by a distinct change in direction. The spaceship, in this case of the "Twins Paradox" as put forth, will actually change course and return back to the earth not the earth turning to go to the traveler.

Thus if the traveler's speed and distance of travel was such that at the end of the trip when the traveler returned, earth time would have progressed by 40 years while time on the spaceship would have progressed by 20 years. Than for the traveler the interval out would be 10 years and the interval back would be 10 years. If the traveler had been 20 years old when the trip started, he would be 30 years old at the half waypoint and 40 years old upon his return to earth. Meanwhile the earthbound twin (the sister) would have been 20 years old when her brother left, 40 years old when her brother was at the half waypoint in her trip and 60 years old when her brother returned. This apparently makes sense and seems to be consistent with observable phenomena and experimentation. Various experiments performed to date indicate a slight slowing (very slight but detectible with accurate instrumentation) of time for clocks that experience high velocity (still not relativistic speed) travel. Based on this information, we should now feel comfortable with the generally accepted answer to the Twins Paradox. Right? Let us return to our scenario with the spaceship traveling at near relativistic speeds, to explore the answer.

We will plot a course for our spaceship that will have it go out in a straight line until it reaches its destination, then return along a parallel course back to earth. The spaceship will perform a 180 degree change in course at its destination (midway point), clearly identifying it as the "traveler". If our course and speed is carefully plotted, the midway point will be exactly halfway in both distance and time for our traveler's total trip. We can then assume that at this point in time, our traveler will have aged one-half the amount she will age over the entire trip. Now let us say that one objective of our mission is to confirm Einstein's principle of General Relativity that states, if you travel far enough through the universe in a straight line you will return to the point at which you started. We will thus plot our midway point for this trip to be the earth, or the same location from where we started.

It is painfully apparent that we have now elevated our Twin Paradox to much higher plateau. Let us look a little closer at our problem. As our spaceship leaves with the twin in the spaceship, as the traveler, should be aging less rapidly then the twin left on the earth. At the midway point of a symmetrically plotted round trip course, the space twin should have aged one-half of the total he will age during the trip and the twin on the earth would have died eons ago. Correct? Wrong! According to our accepted definition of a traveler, it is impossible at this point to tell who the traveler is. For neither the earth or the spaceship has had a change in course.

According to the principles of Relativity, the movements of the two frames of reference up until this point are uniformly relative to each other. That is the twin on the earth would observe the clock in the spaceship to run slower in comparison to her and the twin on the spaceship would observe the clock on the earth to run slower than his. The net effect would be that the observations would have canceled each other out and both twins, if still alive, would be the same age. This is but one example of how the "Twins Paradox" can be validated fully without violating any of the actual or implied rules of Relativity. It should be also noted that the Change in Direction Criteria is not called for in "Theory of Relativity" but is a criteria that was latter implied upon the theory to remedy the problem of the "Twins Paradox".

We could take the change-in-direction criteria, one more step however. We could state that the effect of slowing down the clock does not occur until the change in direction has been made. This I presume could be a perfectly satisfactory solution. This explanation leaves me with a question however. How is it one can live long enough to make the change in direction a hundred-plus light-years away?

There is an even more interesting argument about the Twins Paradox, which is based on the principle of Relativity versus Reality. The argument dismissing the Twins Paradox is based on the traveler having a change in course. On the other hand, more fundamentally stated, in space-time a trip with one or more changes in course is shorter than one in which no change in course has occurred. Now let us look at the Twins Paradox as if we were in actuality planning this theoretical trip. As we start to plot our course, we will need to know the direction of travel (the midpoint that can be any star), the speed of the vehicle, and the point of return (the earth). This is where the argument of course change runs into real trouble with reality. For over the period of time that the spaceship will be gone, the earth will have changed course numerous times while the spaceship has only changed course once. Not only will the earth as a whole changed course but the spot on the earth from which our observer is making his observations from will have changed direction even more times. So, based on the change-indirection rule the observer on the earth is more of a traveler then the traveler in the spaceship. One may argue that the earth is merely orbiting the sun or rotating on its axis and that such angular velocity is not applicable. However a careful study of the spaceships proposed course will clearly show that it too has an "orbital velocity" and what is applicable to one frame of must be consistently applied to all.

In light of this apparent and completely logical validation of the Twins Paradox, we will find that in reality the twin in the spaceship will be the one to age more slowly. Reality is the key to the solution however, and we will see that the change-in-direction rule is not valid. It is not a change in course that distinguishes a traveler from a non-traveler but an even more fascinating reality, which we will explore in more detail as we proceed.

Since this book was first drafted, I have had the opportunity explore several other cases of the slowing clock phenomena addressed in the Twins Paradox in more detail. Including the Velocity Time Dilation, tested by Ives and Stilwell (1938, 1941), and Rossi and Hall (1941), as well as, Gravitational Time Dilation, tested by Robert Pound and Rebka in (1951) and Hafele and Keating, (1971). Additionally, I was associated with GPS system development by the Air Force in the early 1980s and Department of Transportation systems in the in 1990s, in support of other systems I was developing. It is recognized that evaluations of GPS clocks confirm the apparent effect of slowing clocks. What we will try to answer within this work is what is really causing these clocks to slow down.

Chapter 4: Matter, Energy, and Force

The year 1945, the place Alamogordo, N.M., the Law of the Conservation of Matter was to be changed forever. With the detonation of the first man-made atomic device, the true relationship between matter and energy was fully and awesomely exposed to us. What had been a cornerstone of modern physics to that time was smashed. Matter could not only be changed in form but it could be annihilated, as well as, by being turned into energy. We have only now begun to comprehend fully the meaning of this radical change in the constitution of physics. Today some 40-plus years later we are still learning to apply the principles, whose foundations were laid down in 1945. One of the most basic proofs that came from that event, was the full verification of $E=MC^2$. For the first time man had been responsible for the transformation of a significant amount of matter into pure energy.

We now look at energy as an alternate state of matter and we have become quite accustom to the idea that matter can exist in not only different forms but in different states, as well. What we have not fully accepted nor exploited is the reality of this relationship. If matter can be converted into energy and as we have now done, energy can be converted into matter, then energy must exist in a free state. That is that energy must exist, in its own right, just as matter does. When we create matter from energy, where does the energy come from? If energy does not exist in a free state, where does the energy created from annihilation of matter go?

Sure, when an atomic bomb is detonated we see the flash of light or maybe feel the effects of radiant heat but these are effects caused by the creation of the energy. We must still address the issue of where the energy goes when the matter is converted to it. I believe that the physical existence of energy as a component of the fabric of the universe is the real point of scientific enlightenment that we were given by this event on predawn hours of July 16, 1945. The question then is whether we as the scientific community have become so entangled in our complexities that we have become blinded to the simplest truths around us, by our own manmade matrix of convolution.

I think one of the hardest things in this book to agree with at this point will be the essence of this chapter. In short, you will be presented with the simple fact that the universe is full of energy and it exists as little globules of pure energy that we often refer to as photons. These individual globules of energy are smaller than the smallest particles of matter and fill all the space in the universe that is not occupied by atomic sized particles of matter. Within a single atom are almost countless globules/photons of energy. The densest metals are no more of a barrier to these energy globules of energy then a piece of cheesecloth is to a spring breeze.

Imagine a piece of solid gold foil as dense and massive as it is for its slight thickness, which protects our spaceships radiation in space being completely transparent to these minuscule globules of energy that can pass right through it. Impossible you may think. First you may be saying that Newton's concept of universal ether has been repeatedly disproven, and secondly gold is too dense to allow even radiation to penetrate it so it would be impossible for these postulated globules to get through its matrix of densely packed atoms. Yet not only is it possible but as early as the mid-1980s I theorize that it was possible to make a refractive lens out of gold.

It was reported in 2012 it was reported that researchers at the Max Planck Institute of Quantum Optics in Garching along with their colleagues at the Ludwin-Maximilians-University in Munich were successful in using lenses made completely from gold were able to refractively focus gamma rays. A previously impossible action based on prevailing theoretical physics (Domain-b.com, 2012).

The basis for this argument and the explanation of the phenomena identified above comes from the same source, the continuous existence of energy filling in all the space between solid particles of matter. Let us look at an analogous situation that may help demonstrate how matter and energy can coexists amongst each other in the same universe.

Water and Ice are two forms of the same compound H_2O that can coexist side by side. We will ignore the third state of water, which is water vapor, for the moment. Liquid water and solid ice are the exact same chemical compound and the relationship between them could described by the formula $I=WF^2$ where F^n represents the coefficient of freezing that takes place when water goes from the liquid state to the solid state. This type of relationships is almost identical to that between energy and matter. Both are the same thing in different forms just as water and ice are both water in different forms. Thus, energy like water has to exist continually, in a vast universal ocean that not only fills the spaces between planets but also fills the spaces between the smallest subatomic particles. Moreover, like water and ice, energy and matter each have their own specific densities, which are the inverse of each other. In other words in a region of high matter density the density of energy will be less in areas of little no matter the density of energy will be much higher. Energy and matter cannot occupy the same exact space at the same instant, just as two pieces of matter cannot occupy the same space at the same instant. This fact becomes even more important when we look at practical applications of this theory later.

Wait you say, Michelson and Morley proved that could be no ether in 1887 with the now famous Michelson–Morley experiment, which was designed to measure the changes in the speed of light as the earth moved relative to the ether wind. We will see below how this experiment fails because of the same error experience in all previous and later experiments attempting to prove or disprove the constant speed of light. Let us go back to our water and ice scenario for a moment and see if we can create a demonstration of how these experiments failed.

If you were to have a body of water and in that body of water was floating an ice cube. Now the only way that you can sense any disturbances in this body of water is through a sensor that is based on basic fishing bobber technology. Whenever a wave hits the bobber, it signals you that a wave has hit it. You have no other means of detecting these waves, as you cannot see them or hear them. You now measure the time it takes from when movement of your ice cube excites a disturbance in the water sending out a wave propagating across the surface of the water. At some time x after the instant of propagation, the first wave front hits the bobber and causes it to begin oscillating. When the distance between the points in the water where the ice cube was located is measured and divided by the time we will find the speed of the wave to be traveling between the points at a fixed speed we will call F.

Now let us modify the experiment a little and this time instead of being located at the bobber as we were in the previous trial we will back away from it, hovering over the water, and then approach the bobber at some speed less then F as the ice cube oscillates once to create a disturbance on the water surface. Based on the principals formalized in Newtonian physics we should find that the speed of wave front approaching us should be that of the propagation speed of a wave in water added to our speed relative to the wave front. When we get to the bobber however, our only indicator to tell us there is a wave on the water surface we find that no matter how fast or slow we go, the speed of the wave front relative to us is always measured to be the same.

What if we accelerate the ice cube and then oscillate the ice cube in the water so as to produce a disturbance and then measure the speed of the wave front? Again, when we measure the time it takes for the wave front to get from the point of the disturbance to our detector bobber, we find that the speed of the wave front is F, regardless of how fast the ice cube was accelerating or decelerating. Further, no movement changed the relative speed of the wave front to us, no matter if we were moving, the ice cube was moving, or if we were both moving. Does this mean that nothing can exceed the speed of a wave front in water, repeatedly measured as F in our experiments?

Because we as readers are not locked in to the frame-of-reference of our avatar in the experiment described above. We can see the whole event happening and optically measure the relative speeds of all the events taking place. However, our observer, our avatar counterpart, who is locked within our experimental frame-of-reference is limited by the rules imposed on that frame-of-reference and therefore can only detect the wave front in the experiment when it hits the bobber. In other words, the observer cannot see with light or any other electromagnetic radiation the speed of the wave front. The observer being limited by the medium being measured, to measure wave propagation in that medium, can only be deduce that there is no medium and that the apparent propagations of a wave detected with their bobber device are caused by the ice cube but travel absent a medium to propagate in.

What I propose is that prior experiments, such as those done by Michelson and Morley, are flawed in that they all depend on using light or other electromagnetic waves as the detecting medium to verify the presence of the medium they are purported to travel in. The existence of this same flaw in other previous experiments to verify the presences of ether, which is really energy and not any size of matter in its solid form, is enough to reopen the case for further investigation. Yet we are faced with justifiable reasons for reevaluating our view of the universe, such as the question of dark matter and dark energy (NASA, 2012), which was put forth in the early 1990s. NASA published this research about 10 years after the theories in this book were first presented in papers by this author, at Harvard University and Fitchburg State College where this author was taking astronomy, relativity, formal logic, cosmology and other related classes.

Today we have not yet been able to identify either dark matter or dark energy; even though theoretical physics says, it must be there. The reason is we are looking for the wrong thing. The missing mass in the universe is actually the energy Einstein identified in his formula E=MC2. Energy does not appear out of nowhere to make matter, as it exists all around us, like an ocean of water around an iceberg, only it is an ocean (universe) of energy. Consequently when matter is broken down in a chemical or nuclear reaction it releases energy back into the energy ocean, just as a melting iceberg would. A problem that has continually plagued researchers, is trying to isolate and tag energy as a form of ether. Because per the agreed to rules of our frame of reference, which is defining our current state of science, we have nothing faster to use to measure with then light, and light is a wave propagation in this energy ocean that surrounds us.

Understandably you may doubt that this theory has merit, yet with the stars, planets, and interstellar materials in the universe only making up about 4% of the mass of the universe, there has to be something else out there that accounts for the rest of the mass. So is it possible for this energy/ether concept to account for the additional dark mass in the universe. Absolutely, energy is the only explanation for this invisible mass that that has continued to be so elusive. Think about this for a movement, if energy is not being disturbed by some form of oscillation, then it doesn't have any wave propagating (electromagnet radiation) through it and if

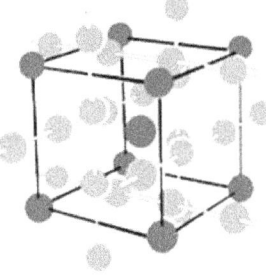 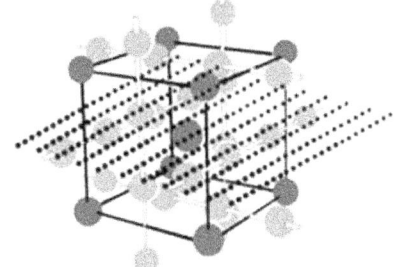

The figure one the left is a representation of the crystal structure of matter in an A15 crystal structure, it represents how we classically think of a material matrix. Atoms are organized in a matrix and the space around and between them are empty. The figure on the right is a representation of the same structure in our ocean universe. The small dots represent the globules of energy that fill the universe in all places where matter particles at the atomic level do not. A15 crystal structure of the A_3B intermetallic compounds (The McGraw-Hill Companies, 2002)

doesn't have waves propagating though it then it will be invisible by our all our instruments that are based on detecting electromagnetic radiation in the form of waves.

As seen in the above figure of the A15 compound, matter is virtually wide open at the atomic level. Further, the matrix even at the subatomic level matter is still wide open space as seen in the representation of an atom in the figure above. When we discuss potential experiments to prove this theory we delve deeper in to how interactions between matter and energy in their respective states occur.

What is an interesting characteristic of this configuration of the subatomic world is that it will easily explain the existence of the dark bands seen in the absorption spectrum of atoms, and the corresponding bright lines in the emission spectrum of atoms. We will explore this concept a further in the chapter on light.

If using a standard tool of theoretical physics, the assumption, we assume that both physical matter and physical energy both coexist in the universe at the same time; we are then faced with the creation of force. What is force? Force is the results of energy and matter trying to occupy the same space at the same time and not being able to do so. Just as our iceberg displaces water in the ocean, matter displaces energy in our energy ocean universe. The real difference being that matter is displacing energy on the subatomic level. Remember earlier how we spoke of computer programs doing functions that have the apparent effect of violating the basic principles of computing, because they combine multiple applications of simple additive functions at the machine level. Thus by obeying the laws at the micro level, one can have an apparent violation of the laws at the more abstract macro level. The same applies here with the interaction between matter and energy, as all displacement of energy by matter occurs at the subatomic level, it is how these occurrences are grouped and stacked that allows for the various phenomena that we see today.

As the subatomic particles of matter are considerably larger than the individual globules of energy (photons), and the globules of energy are considerably more numerous then subatomic particles it is matter that displaces energy, just as ice displaces water. As stated by Newton in Rule 3 of the his Rules of Reasoning in Philosophy, the law of impermeability, alternately stated no two objects can occupy the same space at the same time (Newton, 1729-1768). This law is often attributed to Wolfgang Pauli, who in 1925 established what has come to be known as Pauli's Exclusion Principal, which states that no two identical fermions, which are particles with a half-integer spin, can occupy the same space at the same time. However, this is really an extension of Newton's Rule on impermeability applied to a case in Quantum Mechanics. It is this displacement and the resistance to it that creates what we come to know as force. What is interesting is that matter is not the only thing that can create this displacement, as energy globules themselves also displace other globules. No two of energy globules/particles can occupy the same space at the time, and thus energy and force seem to become synonymous with each other as energy globule in high density areas push outward in a race to reach a state of equilibrium.

The most fundamental argument against this theory is of course the same one that has been employed for over a century. If there were energy globules then we would be able to see them. Yet today here we are in the 21st Century and our physicists are searching the universe for undetectable dark energy. Many physicists claim this invisible form of energy fills some 70% of empty space and should be denser in empty space then on earth. Yet after spending hundreds of millions of dollars on research and elaborate detecting experiment, along with thousands of untold man-hours, they have yet to detect any solid evidence of dark energy, although all the their theories say it should be there. Sounds a little bit like the proverbial ether that does not exist, what do you think?

It is amazing how difficult it is to see something that you cannot see until it hits you right in the eye. However, if you cannot see the ether, can you detect it by other means? Are there some other secondary detection methods that can be used to detect this slothful vagrant of theoretical foible?

Chapter 5: Light, the Phenomena

Previously we discussed the relationship between energy and matter, the potential that energy exists as an intrinsic element of the universe, and that because of its nature energy may only be detected by secondary means. In this chapter, we will look at one of those potential secondary effects of energy that we can detect and study, light.

It is important to first outline and define known characteristics of light before we go on to use light to define the relationship between classical events and relativistic events. Phenomena such as speed and refraction are fairly well defined by modern physics but as in any discussion ambiguity about key concepts needs to be removed up front, so that the participants are all working from the same understanding of the topics.

The speed light travels at, is probably the best-known characteristic of light and all electromagnetic waves, so for the purposes of our discussions in this chapter we will establish this constant at ~300,000 kilometers per second in a vacuum. As the constant is defined for a specific condition, light moving in a vacuum, we should also define a vacuum for our discussions. A "vacuum" is a given volume of space in which all "Matter" is absent. This is a very important point and its importance will become self-evident later in our discussions. It is of no less importance to establish here that light also travels in slower rates when passing through matter and the density of the matter affects the speed in which light will travel through it. What this means is that light for the most part will travel faster through a volume of air then it will through a volume of glass or for that matter a volume of opaque material such as gold. However, when a ray of light exits a volume of matter and reenters into a vacuum it returns to its constant speed of 300,000 kilometers per second. By now you are wondering why I mentioned gold as a material light would pass through. It has been shown by a team of Japanese scientist that very precise lenses can be made out of gold foil that has a thickness of only several molecules. Some thoughts on how this occurs will be address in later chapters.

Fixed length of a ray of light is not a well-known fact. We often speak of a beam of light and a ray of light interchangeably but this is not the case. A beam of light is refers to a ray of light as somewhat of an infinite length such as a beam of light seen coming from a search light. Such a beam of light has no constant direction or duration defined for it (within the parameters of the search light being turned on or off). When we are speaking in terms of hundreds of thousands of kilometers traveled in seconds, even durations of a few minutes can seem as infinity. You probably have guessed by now that a ray of light must have at least two well-defined characteristics that of a fixed direction of travel and that of a very definite duration.

It is quite accurate to assume that a ray of light could have a duration of millions of years however, for the purposes of our discussions we will limit the durations used to create our rays to a few seconds or parts of seconds. A ray of light then is quite similar in character to a ray used in trigonometry having a fixed length and a fixed direction, such that a ray of light, one second in duration, has a fixed length of 300,000 kilometers. This is not to say that it will travel 300,000 kilometers and then disappear. What is meant by this is that if you could see the ray in its entirety from a side view, and you were able to measure it, you would find that its length remained constant at 300,000 kilometers as it continued to move through a vacuum.

Refraction is the bending of light by a medium. Light in general, only travels in straight lines. That is, left undisturbed by outside forces, a ray of light will not change trajectory from that of its original course it set out upon when the ray left its point of origin. If we were to place a laser device on a point on a plane and aim the device in such a manner as to cause the laser beam emitted by the device to travel in a path perpendicular to the surface of the plane. The beam would continue to travel in a path

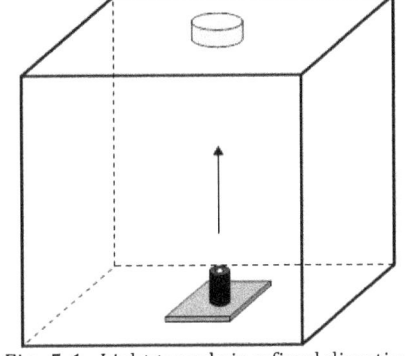

Fig. 5-1: Light travels in a fixed direction and length ray of 1/4 second or 75,000 kilometers.

perpendicular to the plane until acted on by an outside force. Such a force may be the movement of the reference plane or a mirror in the path of the beam. If a common reference point to both the ray of light and the plane is used, then even moving the plane will not affect the trajectory of the ray of light in reference to this common reference point. Absent of course any gravitational variants induced by the mass of the reference plane.

To explain further this concept of light traveling in a straight-line let us imagine a box 300,000 kilometers square. On the bottom of the box, we place our plane with the laser apparatus. We then fire an infinitesimally short ray of light at a target on the top of the box. Moving the plane now, we see that the ray of light still strikes the target. Thus, there is no correlation between the movements of the plane after firing the ray and the trajectory of the ray. Because we are using an infinitesimally short ray, we can further extrapolate this condition to a case in which the ray is fired at the target while the plane and its attached apparatus were in motion (Fig. 5-1). The ray would continue to follow the same trajectory as if the plane was stationary to the reference point and strike the target. If the ray's length was any longer a slightly different event would occur, however that case will be discussed later. We can thus conclude that a ray of light will travel in a straight line when viewed from a reference point stationary to that trajectory.

Next we examine what happens as the stimulation source is put into motion and then a ¼ second ray of light is produced (Fig 5-2). We have estimated the speed of the simulation source to be traveling at about 4 photons per second, perpendicular to the direction of travel of the light

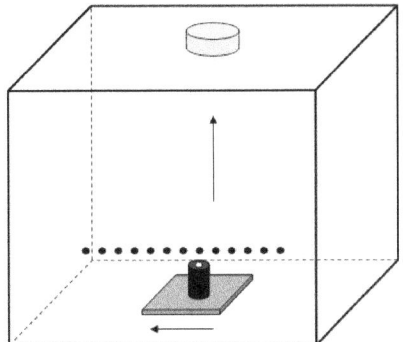

Fig. 5-2: Light ray continues on its original path from origin, as stimulation source moves at the speed of ~four photons per second.

ray. The stimulator fires a round of electromagnetic oscillations while it is adjacent to a single photon and moves on. As the stimulator is at that point for approximately ¼ second it produces a ray of light ¼ second long. Once the stimulator moves on the stimulation has stopped and the light continues to travel in a straight line.

Now imagine if we were to continue the stimulation process as the stimulator was in motion passing by our selected photons. Each of the photons would become a source point for four individual rays of light that travel in parallel paths perpendicular to each of their individual points of origin. Only the most sophisticated sensing equipment would be able to isolate each of these individual rays. Most sensors would register the four independent rays as one continuous ray, with a length of 1 second. Yet each ray actually has a finite beginning and a finite end that can be calculated and registered with the appropriate laboratory equipment.

The above explanation of a ray lays the groundwork for the process in detecting light and being able to demonstrate experimentally that light is actually a wave like propagation in a universal energy medium. In the real universe, light waves would propagate from the excited source photon in a spherical pattern propagating outward in 360 degrees from the source. These waves would interact with waves from other excited sources and can even become stimulators that could stimulate other photons to excited states, allowing them to generate wave patterns. For simplicity sakes in our explanations, we will only be working with that portion of the wave that travels in a direct line (perpendicular) between the excited photon and the detector surface.

The detailed mechanics of how we detect light is another interesting concept, and although not usually directly spoken about in most physics classes, the most important parameter is often alluded to in many experiments and demonstrations. Light cannot

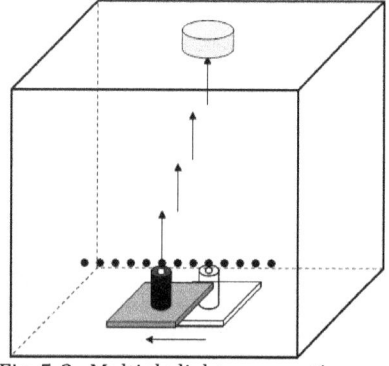

Fig. 5-3: Multiple light rays continue on their own original paths from separate origins, as stimulation source moves at the speed of ~four photons per second past consecutive photons.

be detected by any known method, until it has struck a receptor. Be the receptor an eye, skin, or some mechanical detection devise, the light ray has to actually reach the detector in order for it to be detected. This concept bears some additional explanation.

Suppose for a moment you are standing in a large room and at the far end of that room someone is standing with a flashlight. At a given instance the flashlight is turned on. When will you see the light? You will not see the light until a small fraction of a second has passed.   In other words, you will not see the light until it reaches your eye (we will disregard the additional time required for your eye to notify your brain and for you to realize what you are seeing). You do not see the light not when it departs the flash light or while it is in route to your eye. No matter how close you are to a light source or how far away you are from it, you cannot detect the light until it has reached your eye. As a matter of fact, at this time no means of detecting light prior to its reaching the detector are known.

This concept may be a little hard to fathom, as we are all use to seeing objects before they hit us or we run into them. The great speeds at which light and all electromagnetic waves travel allows us to be able to use them as a detecting medium, however because we are so familiar with this ability in our daily lives we negate it in our scientific reasoning. It is also the reason why we have the apparent perception that the speed of light cannot be exceeded.

Looking at figure 5-4 we see a representation of a ray of light that has been created by the flashlight leaving the light and traveling out the distance of 1/3 second or 100,000 kilometers. At this point, the observer at the sensor has no idea that the light has been created and is now approaching him/her. At this point it is not possible for the observer to measure the relative or actual speed of the light ray as there is no way of the observer knowing that it exists.

A short 1/3 second later the light ray has traveled another 1/3 second or 100,000 kilometers. Now at 2/3 seconds or 200,000 kilometers from the point of origin, at the initial photon that was excited by the flashlights electromagnetic field oscillations. Again the observer has no knowledge that the ray is approaching or ability to detect it and/or measure its speed of approach. Obviously, a common flashlight is not strong enough to travel

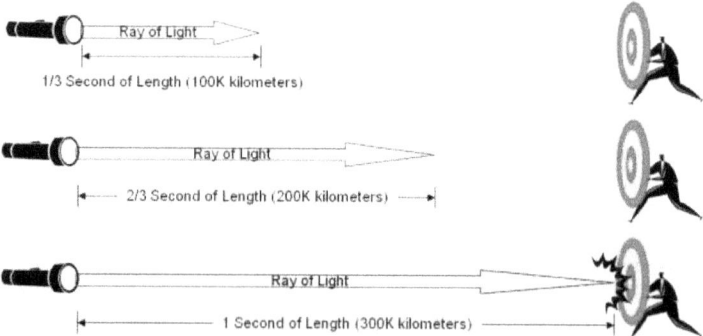

Fig. 5-4: We can only measure that light which we see and we see the light we use to measure it with. Thus the distance between the source and the point first observed is always the distance that light can travel in units of time. Thus distance and speed x time always equal each other.

these distances in a detectable manner. The flashlight for our graphic represents a light source that is both strong enough and resistant to scattering such as laser light source. Note that we are able to measure distance with both seconds and kilometers.

Now 1/3 second later the graphic shows the leading edge of the light ray or wave hitting the target for the first time. The light has now traveled one second or 300,000 kilometers from its source. At this point, the observer has first knowledge of the existence of the light. The observer may be able to see it if it is at the right frequency and intensity and the observers detector can sense the presence of the light ray as it strikes its surface. At this point the observer is able to calculate the speed of the ray. Oddly enough it turns out that the speed is 300,000 kilometers per second.

The most important point to gather from this demonstration is that light cannot be detected until it strikes the observers sensor. You cannot bounce a beam of energy off an oncoming wave front of a ray of light to detect it in advance. Even if you could bounce an energy beam off the wave front, any reflection of the energy beam, as we are able to create and detect, would at best arrive back at the observer at the exact same time as the wave front of the light ray. This is because both electromagnetic waves would be traveling at the same speed from the point where the energy waves were exciting a single common photon. For the purpose of this demonstration, we have limited our ray to one photon in width, although in reality, as we have stated early light normal travels in a spherical wave front from the point (photon) of initial excitement.

We have seen a relatively straightforward demonstration above with a both a stationary source point and target. What if the target was on a moving platform and was headed directly towards the oncoming beam of light at ½ C (velocity of light)? Well according to the prevailing school of thought the observer on the rocket and a third observer able to observe both would see that the light was traveling at C and the rocket with our sensor was traveling at ½ C. According to the prevailing school of thought, since nothing can exceed the speed of light the relative speed between the light beam and the rocket would also be no greater than C. To demonstrate this mathematically one must use what is known as the Lorentz transformation equation, which was deduced in an attempt to rectify the problem raised when all experiments developed at the end of the 19th Century and beginning of the 20th Century seemed to indicate that nothing could exceed the speed of light. Thus, there had to be a mathematical method relating paradoxal observations between two observers in different frames of reference. Because the experiments that were conducted and ordained to determine the speed of light were intricately flawed, thus giving the appearance that nothing could exceed the speed of light, a way of explaining apparent violations of this premise was needed. About a year before Einstein published his theory of relativity, Hendrik Lorentz who believed in the concept of an "ether", was trying to rectify apparent observations concerning the velocity of light with Maxwell's equations. To bridge the two and resolve the apparent paradox Lorentz conclude that one could associate two frames of reference or transform the apparent observations with the following equation, $y = \dfrac{1}{\sqrt{1-v^2}}$. Is this really the mathematical equivalent of a Ptolemy Machine, which is trying to explain the apparent observed phenomena, independent from the actual mechanics of the events, which are being observed?

In figure 5-5, we see a demonstration of an observer moving towards a ray of light being radiated by a stationary source. What becomes painfully apparent is that it makes absolutely no difference how fast or slow the observer is moving towards the approaching light ray when measuring the speed of the light ray's propagation. At T = 0 until just prior to T= 4/6 of a second the light ray is completely undetectable by the observer in the rocket. In the constraints provided by our current science we have no mechanism for detecting light before it reaches us, as the fastest mechanism we have for remote sensing is light (electromagnetic waves) itself. Thus the instant that the light we are measuring the speed of hits our detector, is the instant that defines the point in space from the original source point that we must measure the

65

distant traveled by the light ray in question. As distance and time are interchangeable measures of light, we will find that light always travels at a fixed speed in a given density of energy globules (photons).

To determine speed we divide distant traveled by the time it took to traverse that distance thus giving us units of distance per units of time. In our present demonstration, we are using kilometers divided by seconds. At the instant of time when the light and the observer on the rocket collide, defines the point in space at which to measure distant traveled, and the time duration for the light to have traversed that distance. In our demonstration, this occurs at 4/6 of a second after the light left the source point represented by the flashlight. The point of collision is 200,000 kilometers from the point of origin of the light ray. We divide 200,000 kilometers by 4/6 seconds, normalize the results to a 1 second time span, and we have 300,000 kilometers per second for the speed of our light ray.

Now looking at the closure rate between the light ray and the rocket, we come up with a very interesting result. At the beginning of the demonstration, the rocket and the source point of the light ray are 300,000 kilometers apart. At 4/6 seconds later they have collided, this represents a relative speed of 450,000 kilometers per second. Is it possible that the long-standing postulate of relativity is violated?

For over a century, scientist and physicist have relied on experiments that have all had one fatal flaw common to them. They fail to take into account that we cannot measure light until we see it and at that point, the distance between the origin of the light ray and the wave front of that light ray is that exact distance that light can travel in the time recorded. If you fly a fighter plane

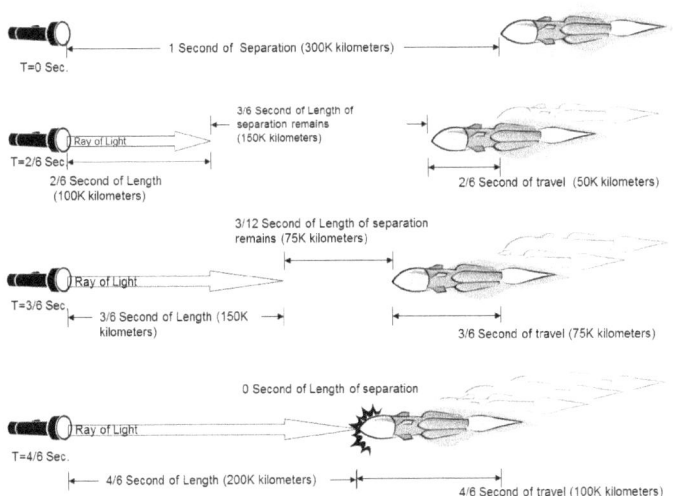

Fig. 5-5: This set of time-delayed figures demonstrates the problem with trying to detect an approaching light beam and also demonstrates the flaw in all previous experiments to measure the speed of light or detect an either-like universal medium. The rocket is traveling at a constant velocity of .5C during the whole demonstration.

from Pearl Harbor to a carrier 360 miles out to sea and you are flying at 360 mile per hour you will arrive at the carrier in one hour. Or after one hour you will have traveled 360 miles. If the carrier was sailing away from you at 20 miles per hour in about 63.54 minutes, you will have arrived at the carrier. If you measure the distance from the Pearl Harbor to where the carrier was when you reached it, it would be about 381.2 miles. Divide the distance from Pearl Harbor, 381 miles by 63.54 and you end up with 360 miles per hour again (rounding errors and head and tail winds are ignored here for the sake of brevity).

Whatever the speed of the aircraft is always the speed of the aircraft no matter what the speed of the aircraft carrier is. Once the plane leaves the runway or deck it becomes independent of the take off point and it does not land on the carrier until it has traveled the full distance to the carrier deck. So using the just takeoff and landing actions, as the start and stop of the experiment, would always reveal the aircraft was flying at the same average speed. If we only had this data, we could assume you could not exceed the speed of the aircraft, yet that is exactly what we do with the data from our speed of light experiments. We can directly relate the light ray to the aircraft in this analogy and we can see that simply measuring the delta between departure and arrival time of the light ray is not a complete representation of what is happening in the experiment. More data is needed, and we will look at how to gather that shortly.

Before we look at a demonstration of an approaching light source, let us briefly revisit the concept of the slowing clocks that we touched on earlier. As we discussed chapter 3 we have experimental and practical evidence to show that clocks slowdown in both situations of acceleration and travel at high velocities, as well as in changing time in response to changes in gravitational fields. The question we will have to answer is that although clocks slow down, does time actually slow down as well? We will look at this phenomena in more detail later but for the purposes of the next demonstration, let us separate the passing of time from the mechanical movement of clock hands (we include the atomic/digital clock as well).

According to the theory of relativity the speed of light cannot be exceeded, therefore an object such as a rocket cannot accelerate to a velocity greater than that of light. However, what if two objects were accelerating towards each other and each reaches a velocity of .65C? According to accepted physics of today, the speed of the two rockets relative to each other would never exceed C, even though simple vector addition would show that the rockets speed relative to each other would be 1.3C. In order to explain this apparent paradox the transformation equations were conceived by Hendrik Lorentz. What these equations do is create a way of relating two frames of reference together in a way that prevents a violation of the rule prohibiting the exceeding the velocity C. But what if the prohibition was apparent, based on flawed experimentation design?

In figure 5-6, we see a time delay sequence of a light source traveling at a target at .5C. For the sake of brevity, we have made some simple assumptions about the demonstration. The rocket was traveling at constant velocity as it passed the time zero point, the figure rounds off the effect of infinite points of excitement and attempts to display just instantaneous snapshots of the unfolding demonstration. Also we are only tracking the original wave front that departed point zero at time zero in our demonstration. Since we can express relative speed by either observer moving or only one observer moving, we can address the issue of both observers moving towards each other, as well as the depicted case of one observer with a light source moving towards the other.

At time zero, a light beam is generated by a device on the rocket, let us say a laser, which is pointed at our observer's target. At time equals 1/12 of a second the wave front of the light beam has traveled about 25,000 kilometers from the point in space the

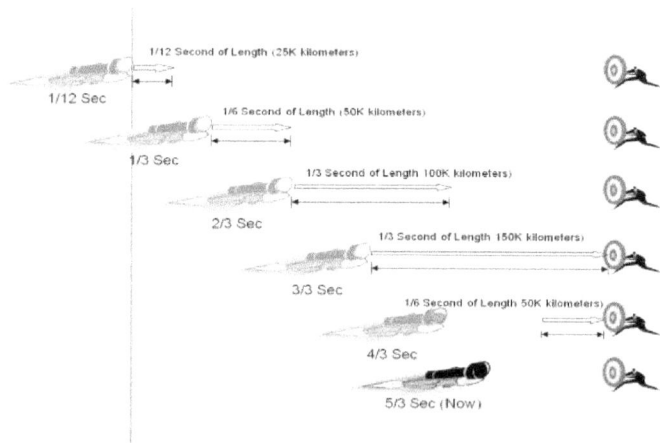

Fig. 5-6: This diagram represents a time delay sequence of a rocket traveling toward at target at a speed of ~.5C, a light is turned on for 1 second during the demonstration.

exciter was at, represented by the zero line in the diagram. At time equals 1/3 of a second the wave front of the light beam has traveled 100,000 kilometers from the zero line in the diagram. The rocket in the same 1/3 of second has only traveled 50,000 miles from the zero line. Freezing time here for a minute, we notice that the wave front at 100,000 kilometers is the same wave front that we previously saw at 25,000 kilometers. Another words the original wave front of our beam of light is traveling from its original source at point zero and not from the current position of the rocket at 1/3 second into the demonstration.

At 2/3 of a second into the demonstration, the wave front is found to be at 200,000 kilometers from the zero point, while the rocket is only at 100,000 kilometers from the zero point. The wave front is continually pulling away from the rocket as it travels independent of the rocket towards its target. If the rocket was to blow up or deviate from its course at this point the original wave front would continue in a straight line (ignoring curvature of space and gravity for discussion sake at this point) towards the observer's target.

At 3/3 of a second, the original wave front hits the target at 300,000 kilometers from the zero point. While the rocket is still 150,000 kilometers from the observer's target. So the observer at this point is not seeing the rocket as it is at 150,000 kilometers away, the observer is only see the wave front that left a point in space (energy ocean) corresponding to the spot the rocket was in at approximately point zero. Since as the demonstration shows, the light wave's and the rocket's movements are independent of each other after the laser projects the light (electromagnetic wave outward).

In most common physics cases and object projected out from a moving platform, is just that projected from the moving platform. Light is different however, light waves propagate outward from a location in space-time that is independent from any material device that creates a disturbance in the either, energy oceans, photon field, or whatever technical term is comfortable to accept. Thus, just like in our previous discussions about creating a wave disturbance in a body of water, the platform we were moving on was independent of the body of water. The wave in the water propagates out from the point in the water that the exciting device, such as a rock or something, hit the water. Do electromagnetic waves propagate from the photon(s) in the medium of photons that was excited by the oscillating device on the moving platform, not from the device and platform?

Thus, we are brought to the concept of phase shifting or Doppler effects. Phase shift or Doppler is another important phenomenon of light. Phase shift occurs when the timing of the arrival of the wave crest of a ray of light are ether compressed together or stretched apart due to the movement of the light source or the apparent movement of the receiver in relation to the oncoming ray. Notice here the significance of the ray being oncoming. A ray of light headed away from you cannot be seen. This is point should be kept separate from the fact that light being emitted from an object moving away from an observer will produce a Doppler effect or phase shift in its spectrum as well. Phase shift is more than an interesting phenomenon; it is a precise attribute of movement in reference to a ray of light, which has many useful purposes in our day-to-day lives.

In figure 5-7, we see a demonstration of how wave fronts from consecutive beams of light emanating from a string of excitement points will pile up one on another over time. Thus, an observer trying to measure the speed of light in this demonstration would again always conclude that light has an absolute speed of 300,000 kilometers per second. The reason being is that each wave front originates from a different point (photon) in space and proceeds outward from that point. This is a much different physics problem than that of say firing a bullet from a gun on a moving plane, where the bullet is already moving at the speed of the airplane and then exits the gun barrel at a velocity of V1+V2. Where V1 is the velocity of the airplane relative to the ground and V2 is the velocity of the bullet relative to the gun muzzle. Thus, V1+V2 would equal the velocity of the bullet relative to the ground as it exits the gun muzzle. Multiple bullets fired would each exit the gun barrel.

The real transformation that is needed when dealing with problems of the velocity of light in physics is transferring from one type of physics scenario to another. An object leaving from a moving platform is one scenario in physics and triggering an independent event located on a separate frame-of-reference as

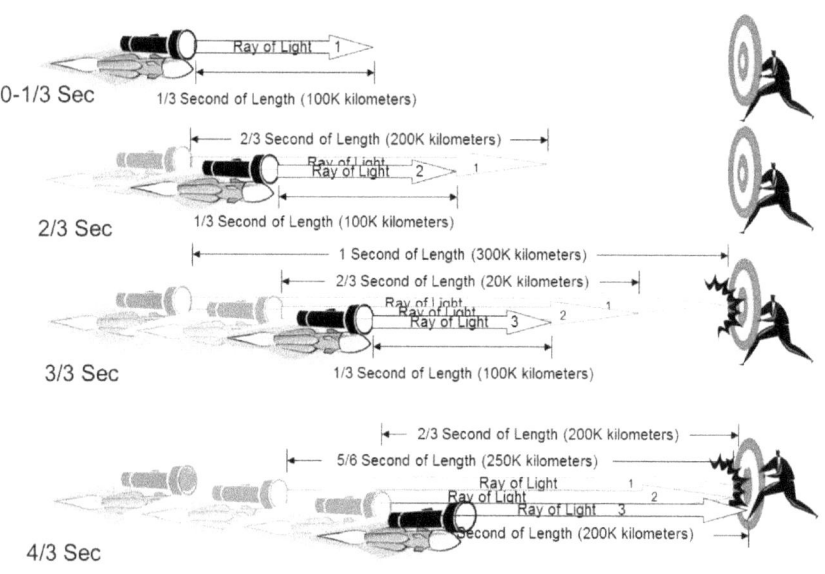

Fig. 5-7: Represents a demonstration showing the phase shifting or Doppler Effect on a light approaching a target when the disturbance source is moving through the energy ocean at a constant velocity of .5C

you pass by it, and is an entirely different scenario in physics. We will come back to this later when we discuss proposed experiments and observations that can be used to verify this hypothesis and to investigate the paradoxes of time in more detail.

Chapter 6: Gravity the Big Push

Release a rock from your out stretched hand and will drop to the ground. If it lands on your foot, it will hit it with enough force to cause considerable pain. Sir Isaac Newton is probably the most famous individual who discussed the nature of gravity, but he was not the first. In ancient Greece Aristotle in the 8th century BC described the effects of the gravity as being a property of the crystalline structure of the universe. While almost a thousand years later, in 628 CE, Brahmashhuta in defense of the heliocentric theory of the universe proposed by Aryabhata a hundred years earlier. In his treatise the Brahmasphuta-Siddhanta, or Book of the Rule of Numbers, "all heavy things are attracted towards the center of the earth" and that "all heavy things fall down to the earth by a law of nature, for it is the nature of the earth to attract and to keep things, as it is the nature of water to flow, that of fire to burn, and that of wind to set in motion... The earth is the only low thing, and seeds always return to it, in whatever direction you may throw them away, and never rise upwards from the earth." (Colebrooke, 1817)

Another great thinker of the medieval world, Al-Kindi, described gravity and was able to describe the variances in gravitational attraction of bodies as a being a property of their distance from each other. Al-Kanzinin wrote the *Book of the Balance of Wisdom*, in which the concept of gravitational forces being dependent on distance between bodies was first proposed. Al-Khazini and another contemporary of his Al-Biruni, also were able to identify the difference between force, mass, and weight. The work of Al-Kanzini offered a basis for the much later theories of gravity that were formulated by Sir Isaac Newton, in the late 17th and early 18th centuries.

In the late 19th and early 20th Centuries, another great mind took up the quest to understand better the forces of gravity and to find a way to both physically and mathematically define them. Albert Einstein began to look at gravity as an integrated part of space and time combined rather than as a standalone force that Newton had previously described it as. In the figure 6-1, we see a comparison of the two systems. Notice that in Newton's version the earth seems to sit on top of a flat universal grid, while in Einstein's version the grid seams to bend under the earth. In Newtonian Mechanics, a gyroscope tends to stay oriented in a constant direction while in Relativity the orientation changes slightly as the gyroscope is affected by the curvature of space-time around the earth.

The question arises as to how two such great thinkers explaining the same observed phenomena come up with such different answers. We are left to conclude that one theory must supplant the other and yet this is not the case as we find both Newtonian Mechanics and Relativity both accurately describe the nature of events within their own frames of reference. At relativistic speeds (high speeds of travel), Einstein's model accurately describes the observed phenomena, yet while at normal speeds, those we would regularly encounter in our everyday lives, Newtonian Mechanics accurately describes the events we observe.

How is it that these two apparently contradictory systems can both be right? How can the universe know when to swap from one system of reference rules to the other as we observe events?

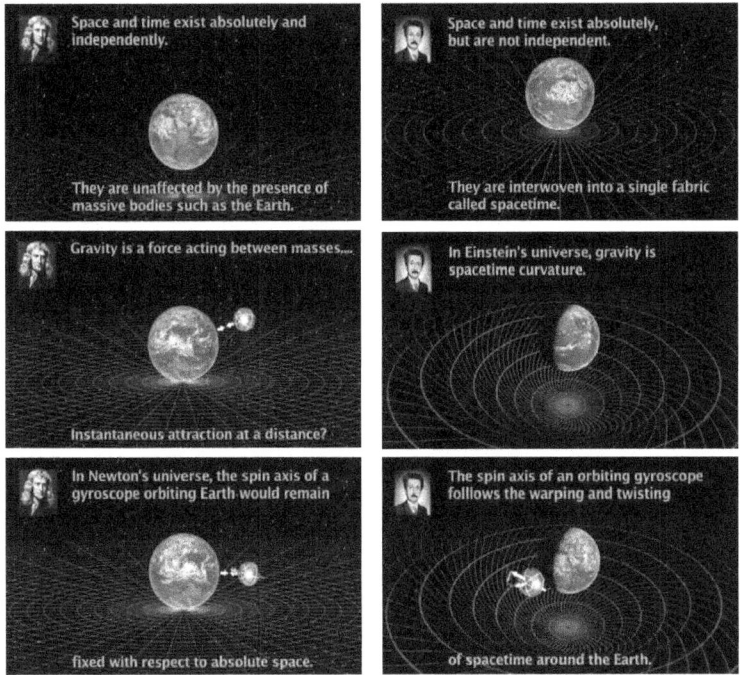

Fig 6-1: In the left column we see a representation of Newton's Universe, while in the right column we see a representation of Einstein's universe. (Overduin, 2007)

Perhaps the answer to the above questions is in a merging of the two theories into a single explanation of the observed gravitational effects. In this chapter, we explain how both Newtonian Mechanics and Relativity are just variations of the same phenomena, which can be more accurately explained, with a single unified understanding of the gravitational force.

According to Einstein's well known formula describing the relationship between energy and matter, $E=MC^2$ we must conclude that energy exists in the universe in some state after matter has been converted to energy and visa-versa that matter is formed from some source of energy that exists in the universe. Traditionally, we have thought of energy more in the sense of a power or force, then a thing of substance, but should we have concluded differently? Suppose for a minute that energy did exist in the universe as a thing, rather than a force and that both energy and matter are the same thing in different forms or states, much as is the case with water and ice. Suppose that energy by its nature consisted of very minute, almost infinitesimally small packets of energy. Imagine that it would take many of these small packets of energy compressed together to make up even the smallest subatomic particle. As even the smallest particle of solid matter would be much larger than an energy particle, and each particle of matter would be displacing a number of particles of energy. If you had one subatomic particle of matter in the universe, the energy particles (photons) would be pushing back against the particle of matter equally from all directions. As all forces would be equalized then the particle would be stationary in the universe.

What happens if you bring into existence a second subatomic particle however? The whole dynamics of our energy ocean changes significantly. The existence of the second subatomic particle upsets the equilibrium of forces on the first particle by creating a path of less force between the two particles. If you are to draw a line perpendicular between the two particles then force being exerted by the energy particles along that line is less then forces of the energy being exerted against the particle from all other directions. As forces always try to reach a state of equilibrium, the subatomic particles will begin to move along the path of least resistance being pushed along by the cumulative forces of the universe. As the move, they will begin to close in on each other and as the distance between them decreases the ratio of force exerted on them along the imaginary perpendicular line between them relative to the cumulative forces of the universe in all other directions will increase until the touch each other. At the point of contact, the force separating them will be zero and the cumulative forces acting in all other directions will be approaching infinity.

Chapter 7: Black Holes and Quasars

One of the most interesting and controversial objects in the universe is one that by name is nothing but by definition is of almost infinitesimally dense matter, a Black Hole.   Karl Schwarzschild a German physicist and astronomer was the first to develop an equation that expressed Einstein's concept of relativity on the gravitational fields of a large mass reduced to an infinitesimally small point. In 1916, Schwarzschild was able to describe what appears to be the reaction of space-time in the vicinity of a massive non-rotating object that has a spherical symmetricity (Schwarzschild, 2008). When one exceeds the limits of what has become to be known as the Schwarzschild radius $R_s = \frac{2Gm}{c^2}$, which is defined by the Tolman–Oppenheimer–Volkoff limit (Kaushal, 1998) or about three solar masses the object collapses upon itself into a singularity that we call a Black Hole (Schwarzschild, 2008).

In the language of the common tongue, if you will, one might reduce this to a simpler explanation of a very dense piece of matter that has been compressed so much by gravity that it has been pressed into an area smaller than say the size of a pin head. The gravity of this very small piece of very dense matter is so strong that once something comes within a certain distance of it the gravitational pull is so strong that it cannot escape it. Even light is caught up in this phenomenon and once it approaches it.

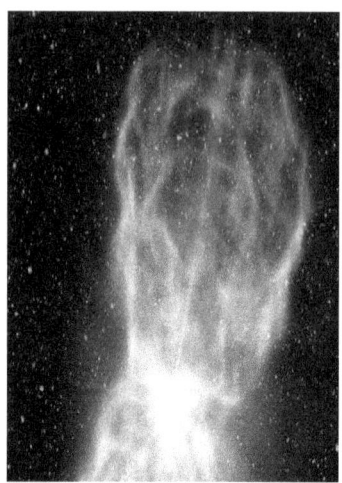

What if there was another explanation? An explanation for the existence of what is described as a black hole, one that has a simpler but more profound explanation for the predicted and observed phenomenon. Is there a way to explain black holes that does not create a new paradox in relativity or require a special case in quantum mechanics?

As we discussed earlier, if light can be described as a wave motion in a medium of infinitesimally small energy particles or photons then perhaps a black hole has a simpler construct. Is it possible that a black hole is simply what its name implies, a hole, a complete and total void in the universe, a place absent of all mater and energy? Quantum physics describes a condition that would be very similar to this bubble of complete void in the ocean universe, often called a false vacuum or perturbative vacuum. In discussing the concept of a quantum vacuum Peter Milonni affirms the zero-point energy state possibility even in the quantum physics, "... *all quantum fields have zero-point energies and vacuum fluctuations"* (Milonni, 1994). Valente describes a quantum vacuum contrasting it to a state of "nothingness" of a classical vacuum, with the state of a quantum vacuum being one in which there is "no (transverse) photons present" (Valente, 2010). If we were to imagine a point of space-time in our mind's eye in which no photons were present, would it not appear as a black hole?

If as we proposed earlier that light is a wave action in a medium of energy particles or photons that are infinitesimally smaller than any particle of physical matter, then if there were an area in the universe that was complete void of these energy particles/photons, "a null electromagnetic field in some region of space" (Valente, 2010), as well as, all forms of matter, that would be a void in space-time, a quantum vacuum, or an area of super nothingness. Although Valente recognizes that the problem of having to maintain a concept of an utter void in space-time his mathematics seems to create a profound paradox in in the applied logic of its outcome, "...the vacuum of the quantized electromagnetic field can be disposed... It seems then that we cannot recover the 'nothingness' of the classical notion..." (Valente, 2010). Perhaps it is the existence of divergence in the perturbative calculations itself that bars the realization of the potential for an area of space-time to be completed void of all things including the very fabric of space-time itself.

Can such a void or bubble of nothingness exist? Consider for a moment an occurrence of the Casimir effect in which the Casimir–Polder forces have been perturbed by an event to conform ``to the constraints of a minimal surface as described by Plateau's Problem. However, unlike traditional minimal surfaces such as a helicoid or a catenoid, we must consider a transformation in which a spherical body can be considered in the set of minimal surfaces. A case for this event can be established by the work of Andrei Moroianu and Sergiu Moroianu on spherical minimal surfaces as describe with a compacted Ricci surface (Vlachos, 1999) with a conical singularity (Moroianu & Moroianu , 2010). If a spherical minimal surface can be created in given volume of space-time the area within the confines of the space would be completely void of both energy and matter. A practical analogy of this state would be that of cavitation produced by propellers in a liquid.

Cavitation is the process of creating voids in a liquid due to pressure differentials occurring when liquids are rapidly pressurized and then depressurized, such as when passed through a propeller or even through a spillway gate at a dam. What is even more interesting, in comparing these cavitation voids to the observed phenomena of black holes, is the similarities with the explosive collapse or implosion of cavitation voids/bubbles in the extreme energy releases seen on the event horizon of black holes. R. Stephen Berry and Boris M. Smirnov present us with an interesting explanation of vacuums/voids in the context of liquids that can serve as perturbation starting point for our explanation of black holes as a void in space-time, "We can thus assume that the voids are simply the relaxed products of vacancies, and that these occur randomly throughout the structure." (Berry & Smirnov, 2005) What Berry and Smirnov have been able to define is that in the transition between two states such as solid to liquid is not always a seamless transition, in that a boundary zone or no man's land that exist between the boundary surfaces of the two states. "The analysis shows a thermodynamically instability of configurational excitations of this system in some range of void concentrations between those corresponding to the mean void concentrations in the solid and liquid aggregate states. This phenomenon makes impossible a continuous transition between the solid and liquid aggregate states that might be induced by variation of the number of voids....The equation for the void flux is given as $j = -(DvN)(dc/dx)+wNc$," (Berry & Smirnov, 2005).

What we have seen in the physical world is an analogy with cavitation voids and state change voids of the observed phenomena of what we call black holes. Of course these analogies only make sense if we are conceding the existence of energy particles that fill the empty spaces between atomic particles within atoms and the vast expanses between atoms themselves.

The conclusion that has to be obtainable is that energy and matter being the same thing (energy) in different states behaves in similar fashions to how matter behaves as it transitions between its various physicals states such as solid and liquid.

We have to ask ourselves if there is a condition of energy as described by $E=MC^2$ that complies with at least the intent of the laws of thermodynamics as they relate to the phenomena of phase transitions. One example of this would be the transition between the SU(2)×U(1) symmetry of the electroweak field into the U(1) symmetry of the present-day electromagnetic field during the initial phases of the early universe. (Lu, 2005) In essence, we have what may be described as the anti-percolation theorem. While percolation theory describes what the distribution of clusters in a random graph would be, by default it also describes the empty spaces in the same graph at the same point in time. If we were to look at this problem of a random graph in the inverse, we could create an argument in which the clusters of dots represented points of emptiness or voids in the graph, i.e. mini black holes that cluster together to form either a singularity or an interlaced cluster.

If this leap of reasoning can be made then we can perpetuate that analogous conditions in the fabric of space-time could result in the formation of similar voids in space-time absent of all energy and matter can exist, as black holes. Note the observation of Lu,

"Just like the liquid-gas transition, it occurs through the bubble-nucleation of low energy liquid phase in the metastable gas phase. If the bubbles are large enough to grow, percolate, they will eventually convert all the gas volume into liquid phase." (Lu, 2005). We have to ask ourselves if it is a form of universal symmetry to look at the gas to liquid transition state as being analogous to the energy-matter transition state that existed in the early universe and continues to exist in isolated pockets that we refer to as black holes. In his paper on Monopole dissociation in the early Universe, Paul Steinhardt, proposes the spontaneous formation of bubbles in the early universe as results of, topological excitations of metastable phases such as non-Abelian monopoles can dissociate and decay through the radial expansion of the excitation core." (Steinhardt, 1981) Steinhardt suggested that this dissociation and decay can reduce the significant formation of bubbles when the transformation/penetration from $SU(4) \times U(1)$ phase to the $SU(3) \times SU(2) \times U(1)$ phase. Thus by default, the reducing from a significant number of bubbles to sum lesser number of bubbles, is an acknowledgement of the creation of some bubbles in the forming in the space-time fabric of the early universe.

An absolute void in the fabric of space-time would be absent of all things including energy particles/photons, and therefore no electromagnetic radiation including light would be able to pass through this zone/bubble of nothingness, giving it the appearance of a totally black hole in space. Although our current state of quantum physics can describe the theoretical existence of bubbles in the universal medium/space-time fabric, quantum theory like Claudius Ptolemy's earth centered mechanical models can only describe the observational characteristics of the physical phenomena not the mechanisms by which they happen. We recall that Ptolemy's model was only meant to show the observational position of the planets relative to an observer on the earth's surface. The model was not intended to map the actual travel paths just project the right position in the sky when viewed from earth. Current endeavors in quantum physics will continue to fail

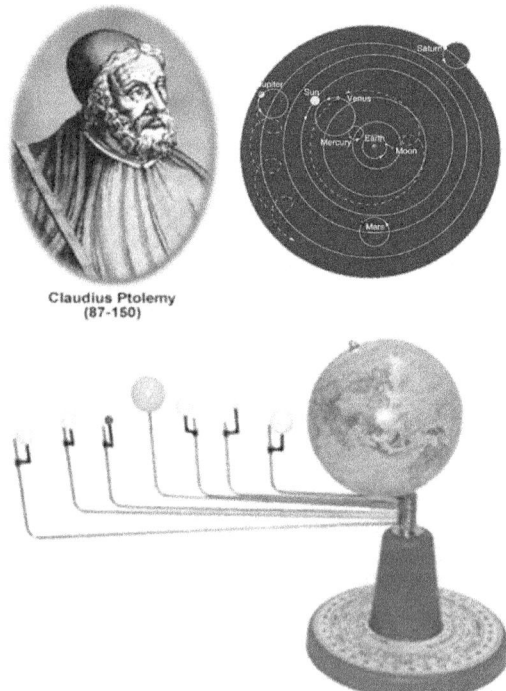

Claudius Ptolemy
(87-150)

Fig. 7-2: Ptolemy's earth centered mechanical model of the Solar System

to accurately and verifiably explain the mechanisms of a black-hole because they are based on observational indicators rather than an repeatable and mathematically verifiable physical reality.

Most of us have experienced a playing with a soap bubble as a youngster or an adult and if you have you share a common experience with the rest of us. Bubbles eventually pop, when an external force acting on the skin creates an imbalance of pressure at a given or the internal pressure exceed the surface tension of the bubble's membrane. You may have even notice the force that the little droplets hit your face after you have popped the bubble and probably accounted that force to the pressure on the inside of the bubble. Well the inside pressure of the bubble is balanced to the outside air pressure or the bubble would keep expanding till it was stretched to thinly or the inside pressure equaled the outside pressure. So where does the force to propel those droplets come from?

Our childhood soap bubble exists in a similar framework as a cavitation bubble except the forces at play are many times, less then what is experienced in a cavitation bubble. Collapsing bubbles can produce bright flashes of light usually lasting only a few milliseconds, this phenomenon is known as sonoluminescence, and the temperature within the collapsing bubble can reach as high as 20,000 degrees Kalvin, four times hotter than the surface of the sun (Physics.Org, 2005). When bubbles collapse they release a large amount of energy, which is well beyond energy levels that would normally be expected. The picture of a collapsing bubble releasing light/energy from Pennicott, 2002 demonstrates the phenomenon. The energy of a collapsing bubble is thought by some to be a form of nuclear fusion. Understanding this phenomenon may help us better understand black holes.

In the context of sonoluminescence of the collapsing bubble, are there events on the cosmic scale that we could relate to as holes, analogous to the phenomenon observed during the collapse of a bubble ?

In the universe we can see an event that is so reminiscent of sonoluminescence that we must conclude is the most reasonable hypothesis to consider next in the continuing research of the phenomena. Quasars are the enigma of the astronomical research. Quasars are thought to be the result of the release of vast amounts of energy when matter is sucked down into a supermassive black hole in center of what is called an Active Galactic Nuclei (AGN).

This process is referred to as an Accretion Process. Most often this process is identified with binary star systems where the more massive of the two stars begins to attract the gaseous emissions of its companion star and as might be expected this process feeds on itself increasing the rate of flow of material from the small gravitational body to the larger over time.

The question arises as to what combination of events and forces could bring about a similar Accretion Process in a galaxy as theorized in a binary star system. One explanation would be a sonoluminescence on a galactic scale. As no one has actually been able to actually see a quasar at a close enough range to describe

Cavatation bubble bursting releasing light and energy.

(Brown & Wren, 2011)

(jinnwe.com, 2014)

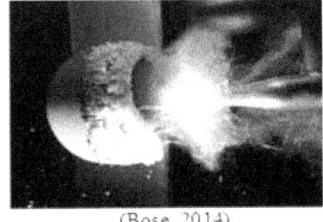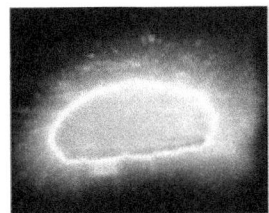

(Bose, 2014)   (Pennicott, 2002)

Figure 7-3:

exactly what is happening, astrophysicist have to rely on mathematical models events and indirect casual-effect observational techniques to formulate a picture of what is happening at these distant events.

However, if we make the assumption as previously stated that it is possible to have bubbles in the fabric of space that contains absolutely nothing no matter and no energy, then we can ask what would happen if that hole, a totally black and absolutely cold hole was to collapse in on itself. When the Mediterranean Sea basin flooded after the Messinian period some 6.5 million years ago, it took some two years to fill with water coming in from the Atlantic Ocean (Garcia-Castellano, et al., 2009). Now the volume of the Mediterranean Sea is some 887.5 thousand cubic miles and it took

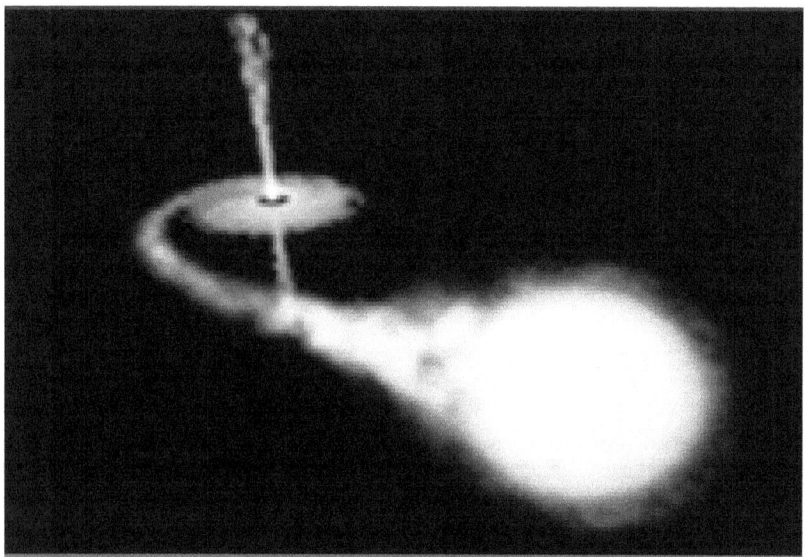

Fig 7-4: Artist's concept of accretion in binary a star system.
(Public Domain)

two years to fill, even a small cosmic bubble could be a millions of times larger than this and could argumentatively take millions of years to fully collapse on itself, which would be the first event in a series of casual-effect events that would occur.

If the energy waves generated by a relatively small cavitation bubble collapsing can generate a bright flash of light; then the brilliance of the light generated by a collapsing cosmic bubble

The Mediterranean Sea took two years to fill when it began to flood.

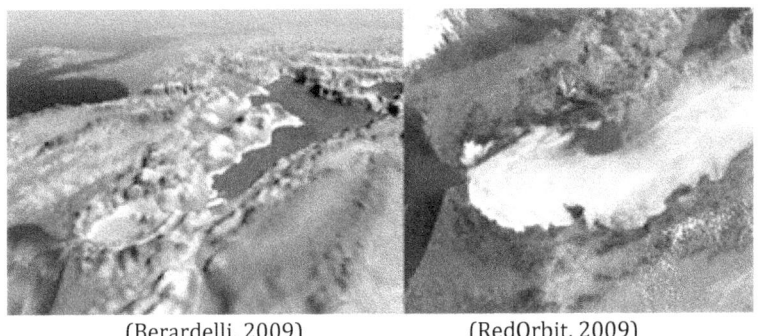

(Berardelli, 2009)   (RedOrbit, 2009)
Fig 7-5:

some trillions of times larger would be infinitely more illumining than a cavitation bubble. Likewise, when a cavitation bubble collapses there are a series of shock waves of ever decreasing intensity sent out from the point of collapse. When a cavitation bubble collapses, it happens in a fraction of a second with an emission time on the order of picoseconds (Brennen, 1995). With an order of magnitude in the trillions of times larger than a cavitation bubble, a cosmic bubble collapse could take place over a period of thousands of years.

Brennen also notes that when cavitation bubble does collapse it does not maintain its spherical shape. Would this aspherical shape in a collapsing cavitation bubble have a similar characteristic in a cosmic bubble? One of the characteristics found often in a Quasar is a pulsating or rotating beam of energy. Is there an analogy to be found in a cavitation bubble? If a cavitation bubble collapses in close enough proximity to a solid surface it will produce a reentrant microjet between the bubble's surface and the solid surface. As time progresses this reentrant microjet will reflect back on the bubble, penetrating the inner bubble and ejecting itself out the opposite side of the bubble from its creation.

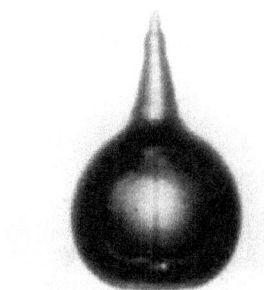

Fig 7-6: Microjet forming on a collapsing cavitation bubble (Brennen, 1995).

W. Lauterborn and H. Bolle describe this phenomenon in a 1975 paper on the subject of cavitation bubble collapse. "This elongation was predicted theoretically by Rattray as early as 1951 and was confirmed experimentally by Benjamin & Ellis in 1966. In the final stage of collapse a pronounced jet is produced, penetrating the bubble towards the boundary as a result of a higher collapse velocity of the upper bubble wall." (Lauterborn & Bolle, 1975). Further Lauterborn and Bolle were able to capture the subsequent events, which provide an even more interesting correlation to our observed phenomenon of Quasars. According to Lauterborn and Bolle, upon the subsequent collapse of the bubble a reentrant microjet forms in the opposite direction of the first microjet. These counter reentrant microjets seem to form from the parts of the bubble that are more spherical in shape as the

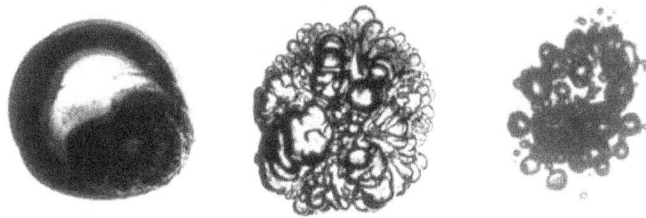

Fig 7-7: A bubble in glycerin demonstrates cavitation bubble collapse (L), rebound/reformation of the bubble (c), and bubbler cloud formation after successive rebounds (Brennen, 1995)

bubble itself elongates (Lauterborn & Bolle, 1975).

Cavitation bubbles also have another property that is similar to the observational properties of a Quasar and Black Holes and that is its spin or angular momentum. As a cavitation bubble collapses and the reentrant jet forms the bubble itself elongates (Lauterborn & Bolle, 1975), the rotational forces begin to redistribute and bubble rotation decreases and increases in order to conserve angular momentum (Kurz, et al., 2008). This rotational forcing is caused by the deforming bubble losing the equilibrium of pressure/displacement against it from the surrounding medium (Choi, et al., 2009). Thus the pulsations we see from a Quasar are analogous to the pulsing that would be seen in a collapsing cavitation bubble as it rotation increases and decreases to maintain its angular momentum.

Just to put this argument into perspective of energy equivalents and time of collapse. If we take an average collapse time of a Xeon cavitation bubble of .625 inches, which is about 69.4 picoseconds (Kurz, et al., 2008) and extrapolate it out linearly to a collapsing cosmic bubble (black hole) of say 1 AU in diameter then the time for the first collapse to complete would be on the order of 654 years. If we assume a modest sized super black hole of about 100 to 200 AUs in diameter then we would have a collapse time of about 65.5 thousand to 131 thousand years. A Xeon cavitation bubble collapse of about 0.625 inches in diameter produces as a flash with energy of about 8.2 pico-Joules (Kurz, et al., 2008) or about 64.2 pico-joules per cubic inch. If we extrapolate that out to a cosmic bubble of say 1 AU then we would get an energy release of about 2.81467E+28 Joules, while an average super black hole (cosmic bubble) 0f 100 to 200 AUs would release about 2.81467E+30 to 5.62935E+30 Joules. Compare this simple extrapolation to that of energy release of a large quasar as reported by the University of California's physics department of about10^{39} Joules observed (UCLA Physics Department, 2014).

Obviously the energy value (energy displaced) of a cosmic bubble is significantly more than that of Xeon but the closeness of the observed apparent energy release to that of a simple linear extrapolation given here is significant enough to warrant further investigation into the hypothesis.

Just as a vortex can form in the water rushing down your drain or in a fast moving stream, currents within the energy filling the ocean universe can interact to create voids or cosmic bubbles of complete nothingness. The surface tension maintained membrane of these cosmic bubbles is such that it is able to maintain a void within because the pressure of the universe around it finds an equilibrium on all points on the bubble's surface. While the bubble integrity is in place no light can pass through its interior because it is void of energy particles (photons), much the same way as sound cannot pass through a vacuum. This lack of energy and the subsequent inability to pass light give a cosmic bubble the appearance of being a black hole. When the pressures on the bubble membrane surface come out of equilibrium, the bubble collapses much as a cavitation bubble collapses or an egg breaks open. The collapse of the bubble is accompanied by a rushing in of energy first in the form of a reentry jet and ultimately after several successive collapses, and rebounds a more sustained wave or tidal flow. It is the shock wave, from the initial collapsing and subsequent rebounds, which release the bright light and energy, we see in the heavens. While the reentry jets are the cause of the pulsar effect coming from various Quasars.

Chapter 8: The Observable Cosmos

If it were possible for material object of cosmic origins to travel at velocities exceeding the speed of light, would they not be apparently visible by astronomers? If astronomers were aware of what to look for then they might have been able to detect such an object in the cosmos, but if they were looking for the wrong characteristics, many such object could exist and be undetected.

Even today after hundreds of years of telescopic observations, astronomers are still learning how little we know about the cosmos. As one anomaly is explained, many more new ones are discovered. Yet with all our science and science fiction computer graphics we still fall far short of being able to explain and or duplicate many of the natural phenomena that we see in the sky.

We generally make the scientific assumption, that no observable objects in the visual universe are exceeding the speed of light. But, are these assumptions based on factual science or are they culmination of a collection of suppositions and consensus arguments in support of vogue theories? Most often we find what we are looking for and overlook what we believe cannot exist. Because of the misunderstanding of the nature of light (electromagnetic radiation) as previously discussed, in it being used as a measure of itself to determine the speed of an object, we need to look for secondary or proxy indicators or an objects actual speed.

One of the first issues that come into play is the concept of an energy pressure front proceeding from an object moving faster than the speed of light. If we remember our previous explanations of our supposition in which light is a wave action in a universal sea of energy particles or photons. A spherical object that is constructed out of a screen like material placed in water would create little displacement relative to the volume enclosed by the spherical structure, so does a body of matter whose atomic and subatomic lattice like structure offer little displacement to energy particles/photons. However, when the sphere made of screen is rotated or propelled through the water at ever greater speeds, it begins to displace more water, until at a certain speed it will displace a volume of water equal to the amount that would be displaced as if it were a completely solid body of the same dimensions sitting at rest within the water. If we increase that speed we will find that the object will create a pressure zone in front of the object in the direction of travel.

As the speed increases this pressure zone increases forward of the object but also it begins to hug and move back along the side of the moving object. Eventually the object will actually displace a total volume of water (actual and induced displacement) greater than the volume of water that would describe the physical parameters of the object if it were solid. This phenomenon leads to two directly observable phenomena that could be used for proxy indicators of the actual speed of the object.

One of these phenomena would be the increased density zone of energy in the energy particle spacing being compressed thus the speed of light (wave action) through the energy will be increased. The second observable phenomenon would be the effects of the harmonic multiplication of the wave intensity as waves are generated and emit outward from within the pressure front zone.

When the pressure zone becomes both dense enough and large enough not only will its effects be mathematically calculable, they will become visually detectable as well. Even though the pressure zone may be limited relative to the immense distances of the universe as a whole, a large object such as the size of our sun could create a sizable pressure zone of several tens of AUs in front of and around it as it approached relativistic speeds. The light traveling through this zone would be traveling a speed much greater than that of light moving the relatively calm universe outside of the zone. Thus, both its measured/observable speed to an observer here on earth would be skewed as well as the perceived distance that the light has traveled. This could also lead an observer here on earth to miscalculate the actual distance of the object from earth or from another known reference point. As the light waves travel over through the pressure zone, they move faster and thus using conventional calculations the actual physical distance they would be calculated to have traveled, would be much shorter than it actually is. This shortening of calculated distance is necessary in order to maintain the classical understanding of hypothesis of the conservation of energy as described by E=MC². For the purposes of this chapter this formula is described here as a hypothesis rather than a law because, assuming our hypotheses of energy and light is correct, in its current use and understanding E=MC² is ambiguous to how it represents the differences of mass versus matter and apparent mass versus actual matter. These ambiguities would need either to be redefined or reproved to be valid in context of the new understanding of energy and light described herein.

Another effect that would be observable is the apparent brightness of an object approaching the observer here on earth would be appear to be much brighter, while the brightness of an object moving away would be considerably dimmer. Let us discuss the approaching object first. As the light waves are generated at the points in the universe where the object is creating them, they will initially speed out from that point, much faster through the denser energy field closer to the object, catching up with those previously generated and not propagating through the ambient energy further ahead of the object. Thus at various harmonics the intensity of the wave is multiplied by these younger faster moving wave fronts catching up to older slower moving wave crests previously created. While this effect only occurs in areas close to the moving object, depending on the size and speed of the object, it could be quite intense. So not only does this compression generate a Doppler effect that increases light frequency it will also make the generating body seem more energetic or brighter when observed here on earth or from another observation point in the direction of travel of the object.

But what if the object is traveling away from the observer, would we have a different observable phenomenon to work with? Just as a when a boat moves quickly through the water or a plane approaches and or exceeds sonic speeds, a clearly identifiable low-pressure zone is created. Even a large object such as a semi-truck moving along a highway can create low-pressure zone behind it, which is clearly observable and physically detectable without any special equipment. When a cosmic object approaches relativistic speeds a similar low zone of low energy density is created behind it and light traveling through this zone would not only be at a lower frequency due to the Doppler effect of the moving excitement source, but also would be traveling at a significantly lower speed and appear to be less bright/energetic. To an unaware observer the reduced brightness/apparent size and speed would indicate a much more distant body, moving away at a slower speed. Another effect would be a pulsing that would occur due to the effect of complex diverging or departing harmonics. This effect is caused as the slower moving light waves of the low density zone are overcome by the inward moving faster waves of the higher density zone as the low density zone behind the object collapses and the two zones converge behind the object to reach the equilibrium state of universal ambient energy density.

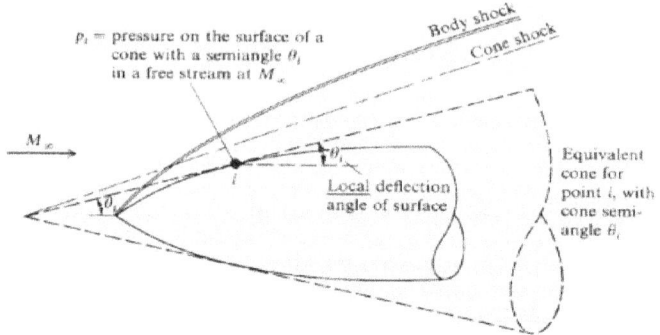

An example of a hypersonic shock wave, showing a complex pressure surface, consisting of a body shock created by the object itself and a secondary cone shock created by the shock cone itself. (Aerospace Web.Org, 2012)

So what would be an observable phenomenon that would indicate that the object in question was moving at a supper-relativistic speed relative to another object and/or an observer here on earth? One possibility for this would come from the conflicting observations of an objects apparent position, direction, and velocity and observed reactions to neighboring bodies. One basis for assuming this hypothesis was previously discussed under gravitational forces. If gravity is thought of as an inherent characteristic of the displacement of energy particles by matter particles then the force of gravity is one that can be thought of as instantaneous. If a gravitational event or change can be measured directly or through a proxy measurement, it will seem to take place simultaneously with the associated event that caused it. Recapping the previous discussion on this hypothesis, if we understand gravity to be a displacement of energy particles (photons) by matter particles then this displacement must take place instantaneously. This concept is extrapolated from the Pauli

Exclusion Principle and the Aristotle Law of Buoyance. From both of these we can easily understand that once a particle of something occupies a space no other particle can occupy that same space at the same time. As mater displaces energy, then this displacement must occur instantaneously or else the matter and energy will be in the same point in space-time at the same instant.

With our understanding of the instant effect of gravity, we can then propose the following. Suppose that an object is approaching an observer at a super-relativistic speed and the object passes within a distance to another observable object as to create an observable gravitational effect. This effect could be a wobble or even pulling on the materials of star or nebula. Because that effect takes place while the moving object is within a limited distance from the stationary one, the time for when the observable gravitational event will occur is fixed by the gravitational effect. As the event is an observable one, then at the instant of the event, a disturbance is made in the energy ocean and waves propagate out from that point at the speed of light. As most of these events happen over time there would be an apparent continuum of events that begin when the object comes with a certain range, increase in intensity as the object reaches its closest point to the fixed object, and then decrease over time until the gravitational effect becomes undetectable. This rise, peak, and fall of gravitational effects can be calculated from the known physical characteristics of the two objects. If however, the moving object is traveling at super-relativistic speeds then it will quickly outdistance the light from the gravitational event and will continue to excite new light waves from the points in space that crosses, which are closer to the observer.

To the observer then the moving object's light that is first seen by the observer, will be that last light generated by the object at its closest distance to the observer. To the observer, the object will appear to be moving away from the observer in the direction from which it came. At a point in time, the observer will see the moving object apparently heading from the observer's vantage point toward the stationary reference object, while the observed gravitational effects will indicate the moving object is approaching the stationary object from the opposite direction headed towards the observer. Additionally if the observer was to look out 180 degrees, a similar looking object would be traveling away from him but would look to be 2 times the distance (we assume a speed S so that the distances and times fit our example) from the position of observation. While the actual distance of the object

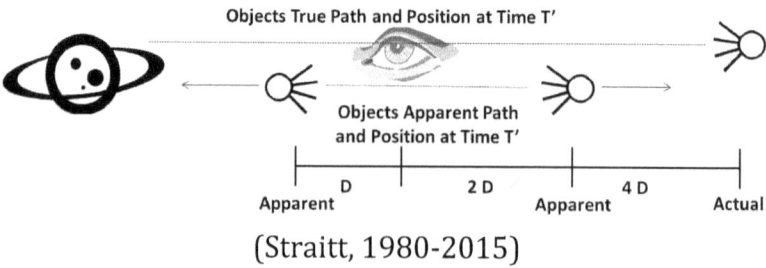

(Straitt, 1980-2015)

from the observer would be 4 times the apparent distance of the viewable object that appears to be headed towards the reference point even though it has actually passed the observer and is headed away in the opposite direction to the rear of the observer. In this scenario, the object at no time will appear to the observer to be approaching the reference object from the far side.

When an object is exceeding the speed of light the light from the closest point of excitement of energy, to the observer, it will always arrive at the observer's eye first. In other words we see the youngest waves first and then than the older waves. We must continually remember that the light's origin is not the object that causes the excitement to induce the wave motion; it is the point in the energy ocean, the place in space where the initial photon was excited, which is the origin of the light beam/wave front. As was discussed earlier, we can model much of lights behavior by thinking of a hovercraft floating over a still body of water dropping little pebbles as it moves along. As would be obvious to anyone doing thi, the movement of the waves in the water is completely independent of the movement or the lack of movement of the hovercraft with you in it. Thus if you dropped a very-very large boulder and created a tsunami, it would be completely possible for you to arrive at the shore line well ahead of the wave and warn the sunbathers to move to safety from a wave they could not yet see.

Up until now, we have only been discussing the most simple of scenarios, very much like a dropping one pebble into a body of water from a hovercraft. You can watch the wave front move out from the point of excitement in all directions forming what appears as a circular wave. In the real universe, the events are much more complicated than this, we must take into account many types of interference waves caused by the event itself, and from other excitement events that are in the proximity. As an example of the complexity involved, one can picture trying to track a radially expanding wave that originates in the bow wave of a large ship in order to grasp the wave complexities that occur in our ocean universe.

SS Cygni is generally described as a binary star system located in the constellation of Cygnus found in the northern hemisphere. The system is believed to consist of a white-dwarf star and a companion red-dwarf companion star (Miller-Jones, et al., 2013). The two stars are separated by about 500,000 Kilometers or just about half-again the distance from the earth to the moon, with an orbital period of about 6.6013 hours (Benson-Avillan, 2010). At this close of a distance, the angular momentum required to maintain separation of these two massive bodies must be tremendous. Imagine two bodies one being about the two-thirds the diameter of our sun and the other being slightly larger than our moon orbiting each other at a distance just slight father then our moon is from earth. Just imagining these two bodies this close together is mind-boggling in and of itself. Imagine the forces that are at work in maintaining the separation of the these two great masses, while at the same time allowing a continuing stream of material to flow from the larger body to the smaller one.

What if we apply the concepts discussed previously herein to the observable phenomena we see in cases such as SS Cygni, would we be able to fashion a reasonable hypotheses that could be proven or disproven mathematically? I believe so, because we have some observable anomalies that although have been addressed to a significant level by the recently announced re-measuring of SS Cygni's distance using radio astronomy in a very long baseline array. What was observed is still open to final analysis and explanation. First, the problem of SS Cygni an apparent binary star system, was that the magnitude of brightness observed was not consistent with the formulas used to describe the anticipated wobble of the paired stars against the distant universe. In 2013, the distance from earth of these paired binaries was reassessed using trigonometric parallax method, which revealed that they were apparently much closer to earth than previously thought.

The distance measured in 2013 show these stars to be some 372 light-years from earth rather than some 520 light-years as previously thought (Miller-Jones, et al., 2013).

Not to suggest that the work of Miller-Jones and colleagues is in error but using this paired system as an example, we could postulate an alternative solution to the problem, if it were possible for an object to travel at relativistic speeds.

Let us suppose one of our members of a binary pair such as SS Cygni was only apparently in close vicinity of its observed apparent partner. If one of the objects were moving at a relativistic speed, we would see a wobble of its partner that does not match the anticipated wobble based on the apparent distances between the two bodies. If a body were traveling at relativistic speeds, meaning that is it is traveling faster than the speed of light, and it was approaching the observer, the light generated by the body would be traveling behind the body itself. If we remember that

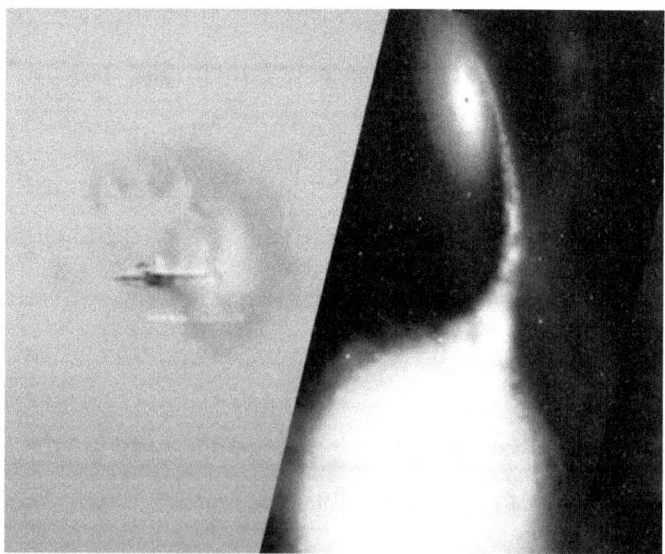

Fighter Jet Sonic Pressure Zone (Rose, 2011) compared to simulation of SS Cygni (CosmosUP, 2014)

light originates from a point in space-time excited by the body, not from the body causing the disturbance, we can understand how the light will be traveling behind the object creating it. Thus, a body traveling at relativistic speeds, will physically effect the second body of a paired system and then move on before the light from either body could even depart the point in space where the interaction of the two bodies occurred.

Imagine it in this manner, as the faster moving body approaches the stationary (relatively stationary for our example) it begins to induce gravitational changes on stationary body. We know from our previous discussions that gravity can be thought of as a form of displacement, such as ice floating in liquid water and thus its effects are instantaneous. Therefore, the stationary body will begin to wobble according to the actual distance between it and the moving body based on these actual gravitational forces. Suppose the moving body is one light unit of distance away from

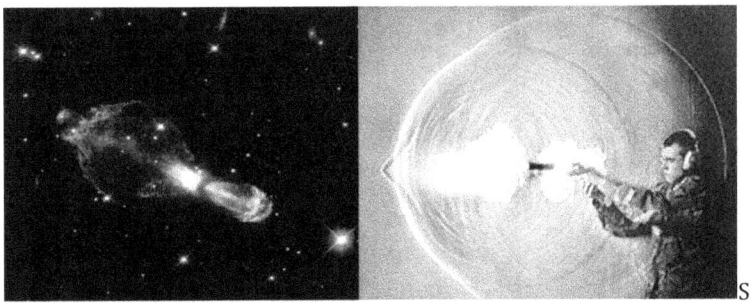

Shockwaves produced by the Calabash Nebula, OH 231.8+4.2 with velocities of up to one-and-a-half million kilometers per hour (ESA & Bujarrabal, 2001), compared to a supersonic shockwaves by a gun firing. (Thompson, 2006)

the stationary body, the body will begin to show the gravitational wobble induced by the moving body approaching it. Yet if the moving body is traveling at one and one half times the speed of light, in just one minute the object will have travelled almost 18 million kilometers ahead of the light it created in time-space 60 seconds earlier.

Let us assume that the point of detectable interaction between the two bodies occurs at a distance of one light unit separation, a year in this case (T_0). When the body moving at relativistic speeds reaches the same distance from us as the stationary body, two thirds of a year (T_1) will have passed and the light from that body will have already traveled about two thirds of a light year closer to us. When two years (T_2) have gone by, the body moving at relativistic speeds will have advanced past the light waves radiated out from it when it was the same distance from us as the stationary body. The moving body will not be at the same distance from us as the first wave fronts of the light sent out from the stationary body one year two years earlier. Assuming that at this point forward in time we could remove the body moving at relativistic speeds from the model, we would find the light radiated from the stationary body at Time T_1 would arrive at our observer's eye/sensor at the same time as the light that radiated from the body moving at relativistic speeds at time T_2. Per our previous discussion we identified that light is a wave function independent of the source of the disturbance that created the wave. Our previous example of a rock dropped into water from a moving platform above the water is an ideal model for this causal effect phenomenon. Thus, the light of the stationary object and the light from the moving object, after it has passed by the stationary object and is much closer to us, will arrive at the observer at the same time. At time T_2 the moving object would have caught up to the light wave front from the stationary object that began at time T_0.

If light, originating from two different times and distances from us, arrives at our eye/sensor at the same time what would the impact be on the analysis of the events we are viewing? Let us break our analysis down into two parts. First, we should address the appearance of the stationary body, both optically and radiometrically. The first observations will indicate the stationary body has begun to wobble with characteristics consistent with being affected gravitationally by a mass approximately one light year's distance from our stationary body. Later analysis would

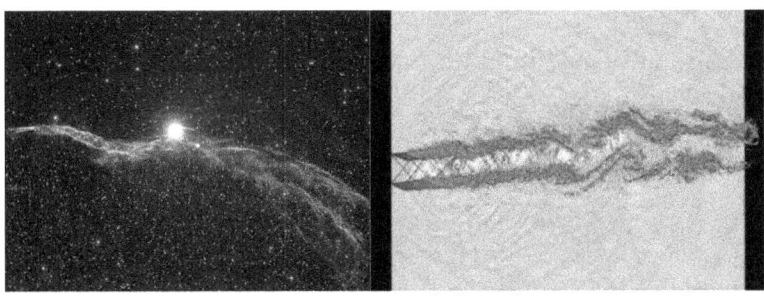

Shockwave of NGC 6960, Western Veil Nebula in the constellation Cygnus (Observatori Ventalló, 2009) compared to the shockwave shown in the acoustic field of a supersonic jet (Schulze & Sesterhenn, 2007).

show that the mystery mass causing this wobble would be equivalent to the apparent mass of our body moving at relativistic speeds. As we would continue to observe the wobble, we would see that the change and the rate of change in the wobble would become inconsistent with a mass that approached from one light year. Rather the rate of change of the wobble would be more consistent with a mass that was approaching our stationary body from two-thirds of a light year and then receding from the stationary body at a similar rate. The sudden changes in or abnormal observations of the wobble may be explained away as errors in measurements or observations, yet something will become more enigmatic as we compare our observations of both bodies.

The second part of our analysis will involve the appearance of the light wave from the body moving at super-relativistic speeds. What we will observe is a pulsing body that appears to be withdrawing from us at less than relativistic speeds. As we continue to observe the body, it will appear to be withdrawing from us and approaching our stationary body being observed. The reason for this is because the super-relativistic body is moving faster than the light waves it is generating in space/time and the last light emitted is the closest to us so it is the first to reach our eye/sensor.

Suppose that both our stationary body and the super-relativistic body were emitting a light flash that lasted for $1/10^{th}$ of one second and then went dark for $1/10^{th}$ of a second at the point of origin. The light coming from the stationary body, would be a ray of light that was about 30,000 kilometers long and would last for $1/10^{th}$ of second followed by a dark period of about $1/10^{th}$ of a second and then another flash of light lasting about $1/10^{th}$ of a second, and so on until the last flash of light from the source arrived at the observer. However, what about the light coming from our moving body traveling at the relativistic speed of 1.5C?

If the light were flashed at the same rate in frequency as the stationary body is flashing we would observe the following. Each flash or ray of light would be about 45,000 kilometers long rather than 30,000 kilometers because the last point of excitement that was at 45,000 kilometers from the first point in space-time. The light would be arriving backwards. That is the last part of the ray to be excited would arrive at our eye/sensor first. If the light ray were carrying a data stream, the message would arrive to us in reverse order. That is "10011101101011" would arrive at our sensor as "11010110111001". The color of the light will have been redshifted due to the artificial lengthening of the waveform over space/time due to the velocity of the disturbing body. That is the wavelength will be longer, as the same amount of force was spread over a greater distance. The interval of time between flashes, or dark periods, will be less than $1/10^{th}$ of second, instead they will be about $1/20^{th}$ of a second in duration. This shortened dark period is caused by the fact that at the $1/10^{th}$ of a second or 45,000 kilometers the body traveled at the end of the first second to the beginning of the third second. The light that was generated at the end of the first $1/10^{th}$ of a second will have traveled and additional 30,000 kilometers or a total of 75,000 kilometers. While the light from the first instant of the 3rd second will begin its travels at about 90,000 kilometers from the origin point, our reference for this scenario. While each pulse of light from our stationary body will last about $1/10^{th}$ second with about a $1/10^{th}$ second dark period, the light from our body moving at relativistic speeds will last about $1/4^{th}$ of a second or .25C (75,000 kilometers) followed by a dark period of $1/20^{th}$ of a second or .05C (15,000 kilometers). Thus, the time for one complete cycle of beginning of one flash to the end of the following dark period is $1/3.33^{rd}$ of a second or .30C (90,000 kilometers). Therefore, the moving body displays an apparent time dilatation of 300% over a similar stationary body. This is an apparent dilation because the length of the light beam is actually shorter.

Another even more interesting scenario arises when the body moving at relativistic speeds is moving away from us instead of coming towards us. In this case, we will still see the duration of the flash of light to be about $1/4^{th}$ of a second, or .25C (75,000 kilometers), but our dark period will be maintained at $1/6.667^{th}$ of a second or .15C (45,000 kilometers). Thus, our cycle time becomes a little less than 1/2 second at 1/2.5[th] of a second in duration or about .4C (120,000 kilometers), or an apparent time dilatation of the actual event of 400%. Compared to the view of the same body approaching us the apparent time dilatation will be 133%. Again, this is an apparent dilation because the length of the light beam is actually longer.

As we said earlier, this chapter is not to suggest that the recent revelations of SS Cygni, which were used in this example, are erroneous. Rather, thoughts laid out in this chapter are meant to spur alternative methods of reasoning that go outside the traditional consensus science approach and ask are there other alternatives that may be able to explain what we observe. Although I will not take the time here to attempt a translation between the two systems, I would venture to assume that in the traditional Relativity Theory and the proposed Ocean Universe theory, the measurements reached by the Miller-Jones' approach would still be the same, as the measurements rely on the observable light effects only. This is because the observable positon of the moving body relative to the stationary body would be would be apparent rather than actual. One would have to know the actual and apparent mass of the object and do an analysis of the wobble of the stationary object to determine the actual vector of the body assumed to be moving at a relativistic speed in excess of C.

Chapter 9: More Cosmic Observations

In the previous chapter, we explored bodies moving at relativistic speeds in excess of C or super-relativistic speed. But what if our relativistic speed was slightly less then C or there was interference taking place between the body moving at super-relativistic speed and one or more other significant light sources or the general background electro-magnetic radiation of the space between the observer and the apparent event observed? Obviously, because of the nature of this book, we cannot get into the complex mathematical model that would be required. However, by limiting our variables/sources of radiation, we can attempt a simplistic explanation of the observed phenomena.

If we have a body moving at a speed slightly under the speed of light and the body was moving toward us we would see a dramatic blue shift in the light coming from the body. The mass of the body would appear to us to be greater than its measurable size and composition would suggest, thus it would have a higher apparent gravity then we would assume for body of its characteristics. This of course is because as a moving solid body in our energy ocean universe would have more displacement of energy particles then the same body that was stationary. An example of this is when we have increased the amount of displaced liquid around a moving object in that liquid, to an amount greater than that same object's displacement of the liquid would be if it were standing still.

Imagine a glass of water filled to the brim and into this glass, you immerse a shaft with a T-bar at the end of it. As you immerse the shaft and T-bar water will spill out over the top of the glass and be collected in the tray you conveniently place beneath the glass. When you get it immersed to the desired depth, you could measure the amount of displaced water collected in the tray you will have the mass of the shaft and T-bar immersed. Now if you begin to spin the shaft at slowing increasing speeds, you will notice more water begins to flow over the top of the glass and collect in the tray. As you spin the shaft faster, more water will continue to overrun the rim of glass. If you stop increasing the speed of the spinning shaft, the water will overflow to a certain point and stop overflowing. When you measure the water in the tray, you will have the apparent mass of the shaft and the T-bar. When you compare your two measurements, at rest and spinning, you will notice that the mass of the water displace by the spinning shaft is more than when the shaft is at rest. Yet the actual mass of the shaft and T-bar remains completely unchanged. The same thing happens when solid bodies move through free space and displace the energy particles that fill the vacuum of solid matter.

As the speeding body continued past us, we would see shift from blue to red, as the waves are no longer being compressed but being spread out by the Doppler Effect caused by the body moving away from us. This can be compared to the Doppler Effect heard when a passing train blows it horn. As the train is approaching, the horn will have an increased pitch do to the compressional effect on the sound waves. This is caused because the train is continually catching up with the previous sound wave. Therefore, the faster the train is going, the higher the pitch will be. Sound does not travel from the train; it travels from the point in the atmosphere where the horn was when the internal reeds, which vibrate the air, were at when the sound wave was created. Thus, when the second wave front is created in the air, the point it is created in is closer to the previous wave front then a whole wavelength. This creates a partial interference happening between waves originating at different points in the atmosphere.

Now as the train passes by you the sound begins to take on a much lower pitch. This is because the opposite affect is taking place as the points of origin of the sequential waves are being spread further apart. If you were to stand on the moving train, while the horn was being blown, it would have a relatively stable sound with a constant pitch being maintained. Likewise the farther you are away and more perpendicular to the direction of

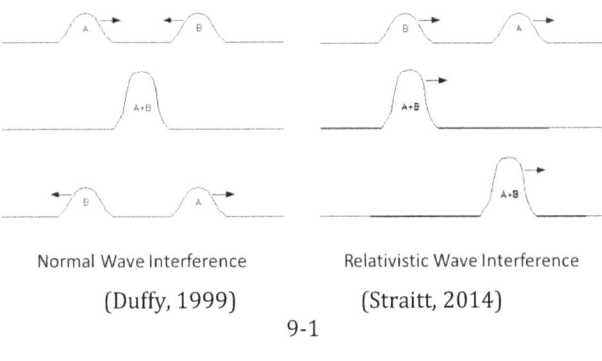

Normal Wave Interference   Relativistic Wave Interference

(Duffy, 1999)   (Straitt, 2014)

9-1

travel of the train the less pronounced the Doppler affect would sound to you. Thus, the body in space creating light waves is analogous to a moving train blowing its horn.

Another interesting phenomenon, "Incidental Interference", can occur with bodies moving through free space at either sub or super relativistic speeds. Incidental Interference, is phenomenon which occurs when a body is moving through space and creates a wave front that forms directly on top of or in perfect phase with another light wave that is traveling on the exact vector that the wave front of our relativistic speeds body creates. When this occurs, the amplitude of the two waves sums and conversely when the waves are 180 degrees out of phase they subtract. Normally this type of increased wave amplitude occurs in nature when we have waves crashing into each other where it is referred to as interference. If the waves are at the same point and if their amplitudes of the two waves either both positive or both negative the wave amplitudes will sum creating a more powerful wave crest. As the waves pass each other, their wave crests/amplitudes withdraw from each other and their individual amplitudes will return to more or less their original state. It would be extremely rare for two waves to be on the same vector and experience extended interference, if one of them is not released from a body moving at a relativistic speed.

In the case of interference when one wave is originating from a disturbance caused by a body, moving at relativistic speeds the inference happens when a new wave is generated over an existing wave exactly at the same point and phase, traveling along an identical vector. The interfering waves will sum their amplitudes but since they are traveling in the same direction at the same speed, the increased amplitude will continue to travel forward until it is stopped by hitting something or runs out of force. If the body traveling at relativistic speeds is moving at a constant speed then a repeating pattern will be created which will osculate from periodic bright peaks as the waves match phase exactly and then to dark periods when the waves are 180 degrees out of phase. Could it be that some of the pulsating bodies that we see in the heavens may actually be relativistic interference caused by bodies moving at relativistic speeds?

When we look out at the universe, what we see is formed by the system of reasoning or logic that we work within. In modern physics, we deal primarily with what is considered a bivalent logic or number system. A bivalent number/logic system has just two possibilities (+) and (-), or true and false. This type of logic/number system became into formal existence with its use by Aristotle, the famous Greek Philosopher from the 4th Century BC as the primary form of logic. In the 19th Century, the mathematical principals of Aristotle became the basis of today's modern mathematics of Formal Logic. In the 1800s, George Boolean formulated a system of logical expression that is today referred to a Boolean-Algebra. This true or false form of reasoning is powerful in solving problems in the abstract world of numbers, but has difficulties in efficiently describing many scenarios found in the physical world. However, there is another type of number/logic system, a system based on three values instead of two. This type of system is used in modern fuzzy-logic computing systems that form the basis of our current computer sciences of artificial intelligence. What if a trivalent form of logic was semantically and numerically applied to the analysis of our universe in the relationship that exist between energy and matter? Would we see the universe in a different light?

First, we need to understand exactly what a trivalent logic system is. A trivalent language is one in which the possibilities are true, false, and maybe/other. The oldest known language that is trivalent is the Aymara language of the indigenous peoples of Bolivian region of South America (Rojas, 1984). Aymaraian is also very unambiguous in its structure and transmittal of ideas. For example in English one might say. "Tom is red haired, unlike Jack who is black haired and he went to the store." Who is it that went to the store? So a trivalent system should have a formal structure that reduces chances of ambiguity and most importantly should have three possible states for any given condition. In contrast to ancient Aymaraian, a modern system of unambiguous language, which incorporates the sense of trivalent operators, is Lojban, which was started by The Logical Language Group in the late 1980s.

One would think that from being able to conceptualize ideas and speak those ideas, speakers of trivalent languages would adopt trivalent mathematical systems, yet this has only partially happened.   Today we attempt to calculate in trivalent logic systems using a bivalent number system, a complex and less than efficient process. What if were to replace our bivalent number system/line with a truly trivalent one? Remember that a bivalent number line has numbers above and below zero. Our bivalent number system can express a quantity, no quantity, or a negative quantity. One has to wonder what a negative quantity really is however. How do you describe holding negative three apples in your hands? We often try to express the concept of negative amount by associating them with debt, such as "I owe him three apples". Logically this is still not a negative. Are you going to give him three apples of negative volume and substance? No, in fact you will eventually pay the debt in three positive apples.   To expand this to our universe, how would we experience, measure, and express negative energy?

One possible solution that presents itself concerning the problem of having negative energy comes from our basic equation of Relativity, $E=MC^2$. If energy and matter both exist at the same time in the universe and neither is really created or destroyed, but

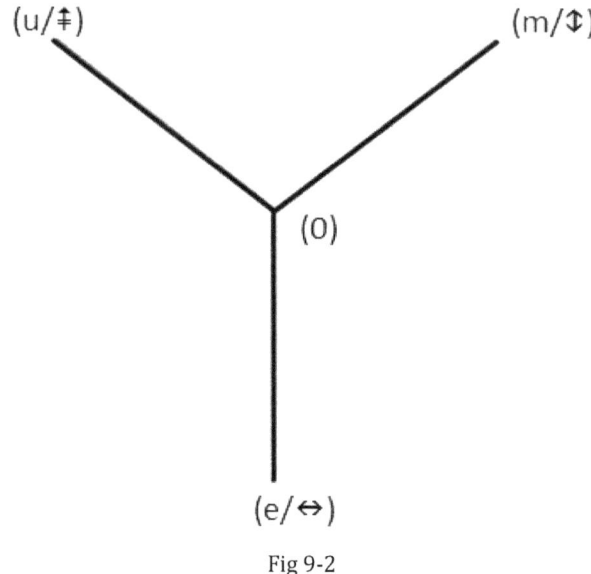

Fig 9-2

transformed from one state to another, then the total value of the universe remains constant, while its components are dispersed proportionally. Thus, we could think of numbers from the perspective of an absolute value branch representing a universal value, a material branch representing matter and analogous to our positive numbers, an energy branch analogous to our negative numbers, and finally we have zero, which represent an absolute void. The power of this conceptual number system is it representation of the relationship between energy, matter, and gravity (universe value) if we let each leg of the number system represent the three constitutes.

In the trivalent number system shown, we see matter represented as 'm' or the symbol ↕. The matter line is analogous to our set of positive numbers. The energy line is represented as 'e' or the symbol ↔. The energy line is analogous to our negative numbers in that it is opposite of the matter line, but it behaves more like our imaginary numbers because the product of negatives multiplying and dividing is a negative not a positive. The universal line is represented by 'u' or the symbol ‡. The universal line is analogous to our concept of absolute values as when 'm' values and 'e' values are calculated together they result in a resultant on the 'u' value line. Understanding that gravity is a displacement of energy allows us to easily see this concept by using the 'u' line to represent gravitational values. The intersect of these three lines is of course zero and within the energy universe zero can only exist in the presence of an absolute void, which unlike a vacuum is devoid of all matter and energy.

The theory of Relativity imposes a limit on the speed at which matter may move through the universe. This limit is based on the proposition that as the speed of an object increase so does it mass, until at the speed of light, the mass of an object becomes infinite. As we discussed earlier this limit is artificial in that what is actually happening is that the apparent displacement of energy particles by the object increases until it latterly displaces all of the energy in a given area of time-space as it passes through it. When viewed in the concept of a trivalent system, the value of the energy component of a given universal value would be zero at that position in time-space but the value of the matter there would be as the same universal value.

With a bilateral logic system contemplating the relationship between matter and energy can be complex and abstract, however in a trivalent logic system the values of energy e $\leftrightarrow$ and matter m $\updownarrow$ on a number line can more readily correlated on a trivalent number line to show how the relationship works.

Let us look at another example of how a trivalent number and logic system could better describe the cosmos that we see around us. Shown in the diagrams 9-3 and 9-4, a door and a three-position switch to control the door, look at the door and try to imagine the knowing where the door is if you are standing there with the switch in your hand and your eyes closed.

If you have the switch turned to the down/closed position and you cannot hear the door moving then you can be confident that the door is closed and you could answer "no or false" to the question of the door being open. If you have the switch turned to the up/open position and you cannot hear the door moving then you can be confident that the door is open, then you could answer "yes or true" to the question of the door being open. However If you have the switch turned to the center position and you cannot hear noise then you cannot be confident that the door is either

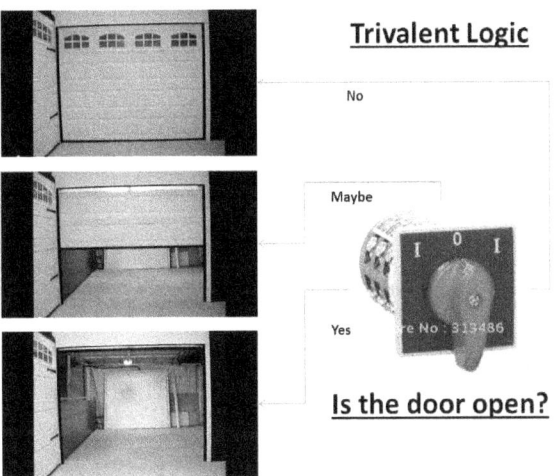

Fig. 9-3

opened or closed and you could not answer "yes or no" to the question of the door being open with any confidence. Although to us familiar with bivalent logic system this may seem ambiguous at first glance, as we think about it we will see that it more accurately describes the situation then bivalent logic.

Using a bivalent logic system, we would not be able to state for sure what our knowledge was about the door. In the example of a bivalent system, we see the switch has only two positions on/move and off/stop. Just because the switch is in the move or stop position it provides no indication of where the door is in the open and close cycle. Now you would be able to determine if it was moving by listening to it move but you could not determine which way it was moving or where the door was in the open and close cycle.

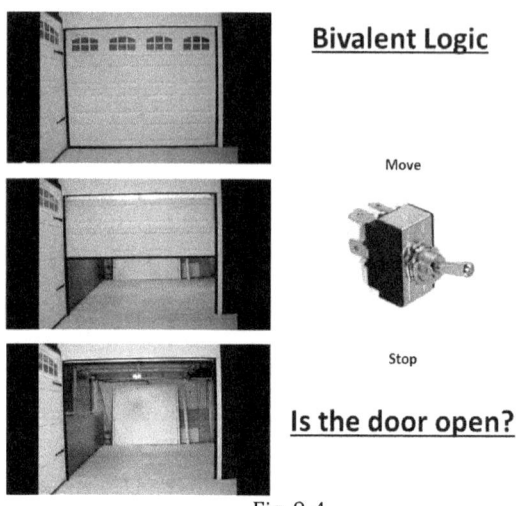

Fig. 9-4

If you hold the switch in the move position until the no more noise is heard you can state that the door has reached the either the full open or the full closed point but not which position. If you let go of the switch while the door is running, so that the switch goes to stop position you cannot say where the door is in the cycle. For example, you may have stopped the door half open/closed, or you may have let go of the switch just as the door had reached the fully closed or fully open position. Without a "maybe" option on the switch our understanding of the world around us is extremely limited and our attempts to describe what we know are ambiguous at best.

A trivalent number system can likewise be defined to remove ambiguity in our exploration of the cosmos by eliminating the confusion between matter, energy, gravity, and the interactions between them. It can also give us a very different way of determining values. In a trivalent number system, the concept of negative numbers as we know them does not exist, thus eliminating ambiguity in describing events.

For example, imagine a block in a round whole and a round dowel that will just fit into that hole. In a bilateral number system we would describe the whole as a "negative space" while describing the dowel as a "positive space". In the figure, we see a round peg and a round whole the peg and the hole are sized so that the peg is an exact diameter fit into the hole perfectly. To define the size of the peg and the whole we will use a common unit such that 1 unit of peg size is exactly equally to the hole size or the inverse of the peg size or -1 unit. Our peg and hole are illustrated in the diagram 9-5.

In the first illustration, we explore a visualization of the mathematics of a bilateral logic/number system to the left and a trilateral numbers system to the right. In the figure on the left, we see that the peg, which is the same depth as the whole and is represented mathematically

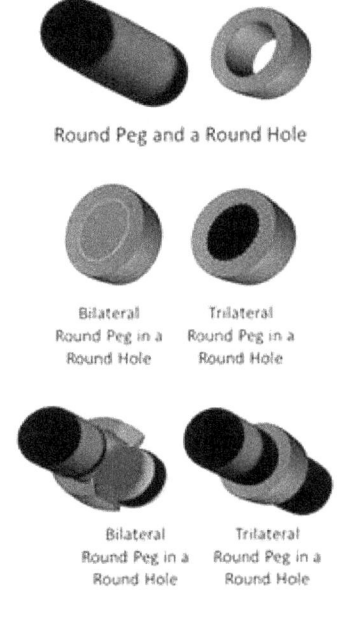

Round Peg and a Round Hole

Bilateral
Round Peg in a
Round Hole

Trilateral
Round Peg in a
Round Hole

Bilateral
Round Peg in a
Round Hole

Trilateral
Round Peg in a
Round Hole

Fig.9-5

as a plus 1, and is completely negated by the whole represented mathematically as a -1. In mathematics, the combination of the peg and the hole becomes zero, or they both cease to exist, at least in the representation of their sums. The dotted circle where the whole and peg use to be is for illustrative purposes and is not part of the mathematics. Without adding additional text to explain, what happens there is no way to deduce the current state from the sum 0 of the formula 1 + -1 = 0.

In figure 9-5, we see a graphical illustration of a mathematical formula describing the same peg in a round hole, using a trivalent logic/number system. The dark circle within the dotted ring represents the value ↕1 unit of peg (+1 in bivalent math) while the dotted ring, which is there for illustrative purposes, shows where the value of ↔1 (-1 in bivalent math) for the hole would be located. The sum value of these two values would be represented as ‡2 or ↕1 + ↔1 = ‡2. The continuing existence of the two independent objects can be represented in mathematical terms directly. The use of the ‡ sign indicates that two types of values are at play in the equation. In the context of our energy universe, we could understand this to be addition of matter and energy to determine an absolute universal value.

Let us look at the other case represented in our figure, where the round peg has the value of 3 rather than 1, while the hole still has a value -1. First, let us do the math, 3 + -1 = 2. So mathematically speaking we have a loss of 1 unit of plug as we are considering the plug as the positive object. The left-hand figure in the bottom set of figures portrays this concept. The bivalent mathematical description of the union of these two objects is that the hole and one unit of the plug have gone to zero or no longer exists. Obviously, we know from physically putting a peg into a round hole; the center portion of the peg and the hole physically does not disappear or become zero values. In order transfer the complete knowledge of the event, we have to go beyond the mathematics and describe verbally the physical state. Generally, in everyday communications we do not take the time to do this, because most of the individuals we communicate with are accustom to this method of communications. This does not change the ambiguity of the mathematical statement as presented in a bivalent context however.

If we look at the example of our 3-unit plug and 1-unit hole from a trivalent context however we get a much clear and less ambiguous understanding of the physical reality without having to provide as much textual clarification. In other words, the mathematical statement will be less ambiguous using a trivalent logic/number system. Looking at the right-hand figure in the bottom set of figures we see that the peg fills the hole with 1 unit and extends on either side by 1 unit of peg. Of course, the extending units of peg could be in any proportion and they would still equal 2 units of peg. Doing the math if we look at this figure with a bivalent number system we would have a formula of 3 + -1 = 2, which indicates that two units have disappeared in total (one + unit and one – unit) or as more commonly thought of 1 unit of peg has disappeared. In trivalent mathematics the formula would be $\updownarrow 3 + \leftrightarrow 1 = \ddagger 4$, which show that we still have the four physical units we started out with.

In the cosmic context, neither energy nor matter can be created or destroyed, then when we package units of energy with units of matter in a given space we need to be able to describe that grouping of units of different elements of the universe. Using an unambiguous system of mathematics and logic is essential to move beyond our current understanding of space and time, as well as, all the objects that exist within it.

To redefine all of the operations of mathematics in terms of a trivalent system would require volumes of books with a small sampling of mathematicians to verify the implementation of all the operations the new math would encompass. The purpose of introducing trivalent mathematics here is to change the perspective and broaden the vision of the reader, enabling you to contemplate the interaction of the various components of our universe in a way that is not generally invoked with our Aristotelian era bivalent mathematics. Therefore, the proof of the various mathematical functions and operators of a trivalent number/logic system are well beyond the scope of this book.

Chapter 10: Energy and Propulsion Systems

One cannot discuss the concepts of energy and universe without discussing the mechanisms involved in moving through this environment. Unless you found a way to step completely out of this reality, then you are moving through the energy universe every minute of every day. As we are all travelers on the spaceship we call earth, which is not under our control, most of us will never have to imagine and/or understand how or why we are moving through the universe. However, what if we were to leave the mothership we call earth and set out on an adventure to another planet or just into space itself, how will we move our crafts through the energy universe?

Perhaps the easy way to begin to understand systems for propelling us across the universe is to understand the systems we use here on earth to propel us through air and water. Surprisingly, although there are dissimilarities there are also many parallelisms between the mechanics for terrestrial propulsion and space propulsion systems that one day may take humans to the far reaches of the galaxy. When we allow ourselves to regard the vastness of space not as an empty void, but as a sea of energy, then we can began to better visualize the dilemma we face with our current state of science when we are trying to travel astronomical distances. Maybe the best place to start is to first look at how we propel ourselves across and/or under the surface of a body of water.

Looking at the most simplest of methods, we would probably start with a paddle. Although a pole to push along the bottom may be considered simpler, we want to address methods here that can work in open water, and not require a physical connection to any land mass. Though our paddle may appear simple, actually operates in two different ways to propel us forward. The first is way is similar to pushing against a wall wearing roller-skates. Newton's third law kicks in, as you push against the wall an equal and opposite forces pushes you away from the wall. When the paddle of a boat is pushed back wards it meets resistance from the water and although the water gives to a great degree it still has enough resistance to invoke Newton's third law. The paddle connected to your arm and ultimately to the boat through your body, conveys the equal and opposite force to the boat pushing it in an opposite direction from the instantaneous point in the water. As the paddle is moving in an un-uniform direction and velocity through the water the boat is constantly changing its vector slightly and/or your body repositions slightly to adjust the direction of the force to maintain the boat on a constant vector.

The velocity at which a paddle moves through the water is significant to its ability to move the boat forward. If a paddle is moved through the water very slowly, the water will part and the amount of displacement will be about equivalent to the amount of water displaced by the paddle standing still in the water. If you move the paddle slow enough through the water there will not be enough force created in the opposite direction to overcome the resistance of water and gravity and the boat will not move. If you look at a slow moving paddle, you will notice there is virtually no displacement-void on the leeward (the side towards the direction of boat travel) side of the paddle. As you move the paddle faster through the water, the first thing you will probably notice is that the boat begins to move forward, but also that the water on the leeward side of the paddle begins to have a lower level then water on the windward side (side opposite the direction of boat travel) and along the edges of the paddle. What is happening is that the water is being displaced by the moving paddle more then by the stationary paddle and the displaced water is mounting up on the windward side of the paddle to exert force in all directions as it tries to reach a state of equilibrium with the water of the lake.

As the displaced water moves outward in all directions, it pushes the paddle forward. If this did not happen, the paddle would just move backwards through the water. As the water cannot be compressed as a gas would, then the resistance to the force exerted by the paddle is transferred to the water and temporarily stored in the increased height of the water on the windward side of the paddle. This is not the only force in action that is acting on the boat. As the water is displaced to the rear of the boat, it creates a higher density of force behind the boat and a lower density of force in front of the boat. To think of it in other terms imagine that as you are paddling the water to the rear of the boat you are creating a wave just like you would see at the ocean with a surfer riding it. In our paddling scenario, your boat is the surfboard and you are the surfer, and of course, the little wave you are creating is not going to cause you to go very fast, never mind create a wipeout. Yet because of the virtual non-compressibility of water, if this displacement did not take place you would not be able to paddle and propellers would not work.

The ability to have a localized increase in the density of force is what allows us to propel objects through a medium. Visualize an enclosed container completely filled with water to the point that there is no free or compressed gas in the container and there is no room for the water to move around in the container. If immersed in the container is a model submarine with a motor and propeller. When we remotely switch on the submarine's motor on what will happen?

In an open body of water like a pool, lake, or ocean the propeller will begin to turn, displacing water from one point to another. The leeward (front/low pressure side) side of the propeller will have a lower area force-density and the windward side (rear/high pressure side) will have a high area force-density. Because we are in an open body of water the volume of area that the water occupies, within the open container holding it, will increase by the amount of the additional displacement caused by the spinning propeller and of course the model submarine will move forward going faster as the propeller turns faster. If our open body of water is being held in a container small enough, then the displacement caused by the propeller spinning will cause the water level in the container to rise by an observable amount. In a large body of water such as a lake or the ocean, the rise is spread over such a large surface area that it becomes unobservable and in most cases even unmeasurable by our current technology.

If you put the same model submarine in an enclosed body of water, such that the container is completely filled and there is no room for the displaced water to expand to and turn on the motor the outcome will be entirely different. While the propeller is turning at an extremely slow rate where the infinitesimally small

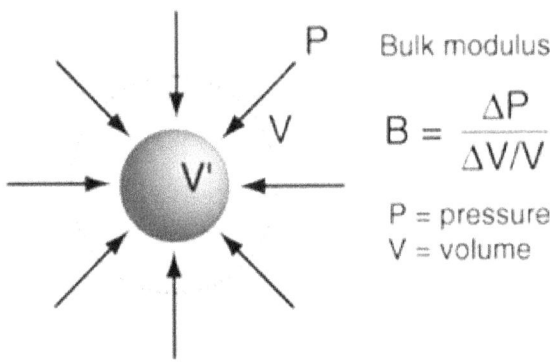

Fig. 10-1: (Nave, 2005)

amount of displacement that would occur, would be absorbed by the minute compressibility of the water, which is known as its "Bulk Elastic Property". For water, this elasticity is calculated by taking the inverse of the "Bulk Modulus" of the liquid, which describes its compressibility as a ratio of the change of pressure to the volume of the liquid. For water the Bulk Modulus is calculated as $B_{water} = 2.2$ x 10^9 N/m^2. According to C. R. Nave, of the University of Georgia, "*At the bottom of the Pacific Ocean at a depth of about 4000 meters, the pressure is about 4 x 10^7 N/m^2. Even under this enormous pressure, the fractional volume compression is only about 1.8%*" (Nave, 2005). What happens then as the propeller speed increases? Once the propeller reaches a rotational velocity, where the volume of the displaced water is greater than the Bulk Modulus, and where the resistance to increasing the compression of the water to its maximum compression level requires more force then the motor can generate, the propeller will stop increasing in speed and/or the motor or some other part of the propulsion system will fail.

So in a sense we can think of paddles, propellers, and other similar propulsion systems as being mechanisms to create a virtual endless wave behind the vessel that the vessel virtually surfs down, just like a surfer on a monster wave at the beach. There is a limit as to how big the virtual wave can become, however. Once a paddle is moving fast enough, or a propeller is spinning fast enough, so that it can maintain a total displacement of all the water in its area of virtual volume, then the water in front of the propeller for example cannot come into the area circumscribed by the spinning propeller, fast enough to enter the space between the blades and be expelled out the other side. At this rotational velocity, the forward propulsion forces would cease and the boat would come to a stop. We will see later in the discussion how this relates to the concept of an objects mass going to infinity as its velocity approaches the speed of light.

Before we leave our boat on the water model or under the water, we need to look at another type of propulsion system, wind power. It is difficult to figure out which came first to our ancient ancestors the idea of the paddle or the idea of the wind blowing a

Fig. 10-2: Wall paintings of Aboriginal Sailing Ships found in Australia dating back to ~50,000 BCE (Strong, 2014)

boat across the water. In either case at some point several tens of thousands of years ago our ancestors discovered how to use sail equipped boats to cross the oceans from Africa to Australia and on to the Americas. Where and how did these people originate the idea of sea travel is yet to be determined.

Although we generally think of sails as a device the catches the wind and transfers the movement of the wind through the sail and mast to the boat, there is also a type of sail that is placed into the water to catch the movement of the water to transfer to the boat. These sea sails or sea anchors as they are often called are generally used to help control a boat in rough seas, they can also be used to propel a boat along in the direction of the current. Of course, the speed of the boat from point A to point B is generally limited by the speed of the current. Another observable factor is that as the boat's velocity increases to the velocity of the body of

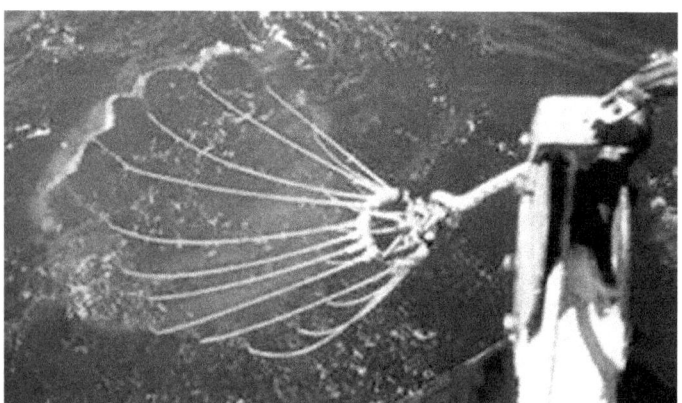

Fig. 10-3: Sea Anchors/Sails have been used for Millenniums keep boat bows into the wind and keep a boat stable in rough seas. (Luard, 2014)

water it resides in, its relative velocity to the body of water approaches zero. This phenomenon bears a more in-depth exploration before we move on.

When a boat sits at anchor, it is generally considered at rest relative to the earth frame-of-reference. If we could look down, to see the water moving by the boat, and assume the boat is moving through the water. In fact if we were to change our reference from the earth to the surface of the water, we could then describe the boat as being in motion and the water stationary. In this state, the boat would be at its maximum velocity going through the water. If we were to take into account the relative changes between just the boat and water, we create a sequencing of events that to us would travel in one direction. In fact, we would have created a timeline that could be used to describe time for all our observers in the reference frame.

If we were to cut our anchor line and continue to observe ourselves relative to the water from the surface of the water reference frame, we would find that our speed continually began to slow down but never quite reach zero. As the changes between our boat and the water surface became less we see this as a slowing down of the rate of change in our timeline. Of course, if we could somehow step out of the reference frame and look in from the outside we would be able to view all the events occurring within the reference frame simultaneously. For the outside observer, time internal to the reference frame would take on an entirely different perspective because at any given time the outside observer could observe any event and all events at the same time. This is something that an internal observer cannot do. We will explore this concept of time in more detail at a later point in this book, but for now, what is most important is the understanding of the difference of motion between an object in the framework of the reference frame and motion between two objects residing within the framework of reference.

If we were to attach a sea anchor/sail to the boat and it was operating at close to 100% efficiency, so that the force on the sail was able to overcome wind drag, gravitational forces, and other forces operating to retard the movement of the boat, we would find the boat would soon come to a complete standstill relative to the water it was in. However, relative to the stationary frame-of-reference observer, such as an observer on the beach, the boat would be moving along at the same exact speed as the water it was suspended in. So depending on your frame-of-reference the boat is either moving or standing still, but in no case will the boat move at a speed exceeding that of the moving water, unless an additional propulsion force is applied to it.

The concept of a sea anchor/sail can be applied to space propulsion as well in the form of what is referred to as a solar-sail. A solar sail works in the very same fashion as its water bound cousin by capturing the force of the moving energy in the universe and being drug along with the current. This force is often called the solar wind, but we will show in our following discussion on wind sails and boats, how the so-called solar wind is analogous to a current in water, rather than wind blowing above the water. A solar sail is bound by the same types of constructs as a sea sail, in that that its ability to propel a load is based on its size and the strength of the current it is in, as well as, its maximum speed being that of the current pushing against it.

Unmanned spacecraft such at the IKAROS space probe launched in 2010 are now using this technology to make their way across the solar system and eventually to coast on into interstellar space. However, would this approach to space propulsion systems be able to move people and equipment to other planets in our solar system as well as to distant solar systems and galaxies? To answer this question, we would have to address two significant issues with transporting humans through space-time and mass.

Time is continually going to be a factor in human space travel because of our short life spans relative to cosmic time scales is infinitely small. Space propulsion systems have to be able to move our crafts across the cosmos at velocities that allow humans to get to their intended destinations within the crews usable lifetime. The maximum effective velocity that a solar sail could propel a spacecraft is C or the speed of light.

To go faster than C it may be possible to adopt a methodology from water sailing known as tacking the sail. Tacking the sail works to increase the speed of a boat beyond the actual speed of the wind that is pushing against the sail by using applied vector analysis. In 2012 a world speed record for a sailing vessel on water was reached by the Vestas Sailrocket, when Paul Larsen was able to pilot the vessel to reach a speed 2.5 times the prevailing wind during the run (Wikipedia, 2015). When a boat is tacking and the true winds on the boats are about 135o to 140o off the direction of travel creating an apparent wind on the sail of about 45°. When this happens the sail goes from being a fiction device that propels from being dragged along to being a flying device that creates lift, which has the apparent of effect of being sucked rapidly into a low-pressure zone on the downwind side of the sail. As the difference and pressure between the two sides of the sail increase, the velocity at which the sail falls into the low-pressure zone increases and the boat moves faster accordingly. The phenomenon that creates this pressure differential or "lift" as it is more commonly called, is a special case of a "Laminar Flow" and which the distance along one edge of the surface creating the Laminar Flow is longer than the distance on the opposite side. When this conditions happens the air on the longer side becomes less dense, which is consistent with effects defined by the Bernoulli Principle. This is the same phenomenon is what creates lift on a plane's wing causing it to fly or on the surface of a wind turbine blade causing it to spin

Just as in ocean sailing, it may be possible to allow the solar wind to hit the solar sail at an angle, while using some technique to maintain a lateral resistance against the fabric of the energy universe to achieve a course of travel that is at an angle to the direction of the current of the solar wind, which would be greater than C. We need to question the viability of this apparent free

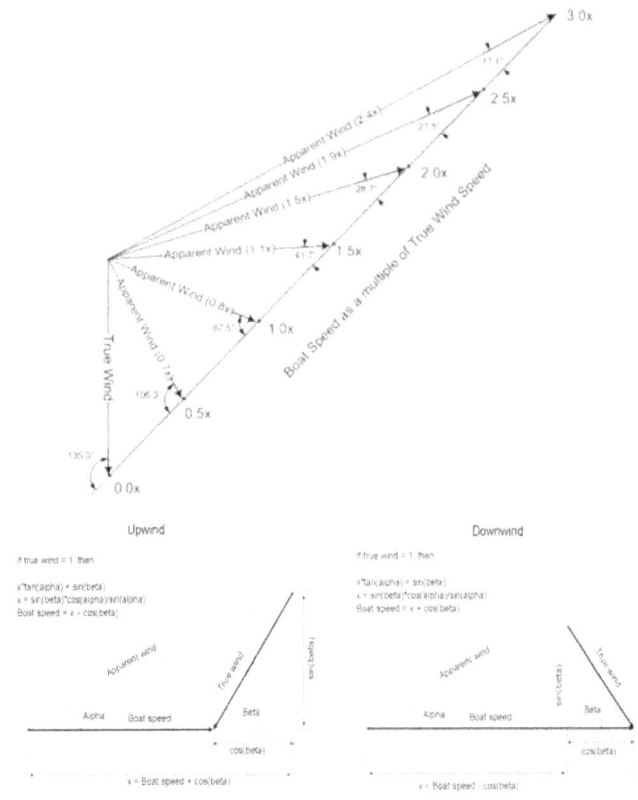

Fig. 10-4: (Wikipedia, 2015)

propulsion source, the solar wind, and its potential ability to propel spaceships at the speed of light or faster if a form of solar sail tacking can be performed.

The reasons we have to question this propulsion system are several folds. First as our actual experience with it shows it requires extremely large sails to collect enough solar wind to be useful in moving manned space-craft. Sails measured in terms of square miles/kilometers would be required to accelerate and maintain a space vehicle to a useful velocity. Tacking a sailboat on the ocean is possible because the water creates a lateral resistance against the hull of the boat keeping the wind from pushing the boat sideways. What would create such a lateral resistance against the hull of a spacecraft? When we think of a wind-jamming sailboat we are looking at a second force pushing the boat through medium water. A solar sail powered spacecraft would act more like a vessel that has deployed a sea-anchor/sail then a wind sail, thus drag rather than lift is the primary propulsion component.

As promising as solar sail may appear to be there are also practical limitations to their use. Once a sail more than a mile long on each side is deployed by automated means, how is it repaired if it becomes damaged? We simply do not have the technology that would allow a human to work in space to repair a torn or ripped solar sail. A solar sail several miles square would be an easy target for meteors and other space debris. How much damage could a solar sail withstand before it became ineffective? What would happen if a large asteroid were to impact the sail and pull it and the space craft into an orbit around the sun or other planet? Even worse, an asteroid in interstellar space may grab a solar sail and send the sail and spacecraft floating through interstellar space indefinitely. Another problem occurs when we finally get close to our destination, and the solar wind from that systems sun becomes stronger than the wind from our sun. How do we propel ourselves towards that solar system, if we are being pushed away from the destination sun, and cannot tack a course towards it?

Although our exploration of solar sails has not been that fruitful in identifying a viable propulsion system for large or

Fig. 10-5: An Artist depiction of the IKAROS space-probe being propelled by a solar sail, which was launched on 21 May 2010 by the Japan Aerospace Exploration Agency (Mirecki, 2011)

critical payloads such as manned space missions, we have further identified evidence to help us understand the nature of our ocean universe. Simply put, if energy did not exist in open space as a real component of that space, it would not be able to create a frictional force on a solar sail and drag it along down current of the energy. It also begs the question as to the possibility of the existence of a force in the universe that would be analogous to the wind across our terrestrial oceans.

Although we have not really identified a viable propulsion system we have identified what could be a common thread in how all propulsions systems work. In each of the examples we have looked at so far, we see that as the force density in a given area increases it tends to want to equalize itself with its surroundings. It does this by spreading out with equal force in all directions from the center of mass of the force. If we apply our understanding of energy from previous chapters here, we can redefine our understanding of force density to actually be an energy density. Such that force, is actually a localized increase the energy density that creates a force as the energy tries to dissipate equally in all directions. Whether we are piling up water behind a boat, catching wind behind a sail, or blasting a ball of fire out of a rocket engine, we are generating a localized region of increased energy density that is constantly trying to equalize itself by spreading out with equal force in all directions.

Many of us in high school or college Physical Science or Physics classes learned the general principles for how a rocket motor works. As the fuel ignites in the combustion chamber pressure builds up equally within the chamber exerting an equal amount of force in all directions against the chamber walls. On the back end of the combustion chamber, is a hole called the throat, which leads to the rocket nozzle and out into the environment. Thus the forces inside the chamber become unbalanced an the rocket moves forward pushed by the forces at the front of the combustion chamber while the exhaust gases go out the rear of the rocket lowering the forces pushing against the rear of the combustion chamber.

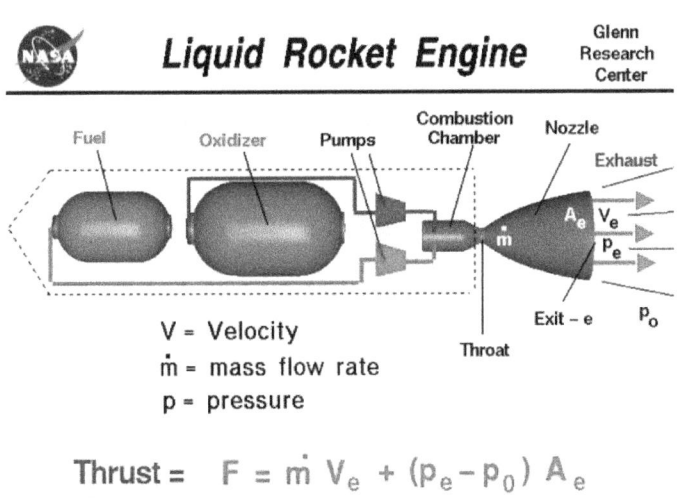

Rock Engine Diagram

Throughout my academic and professional career, the above description of rocket dynamics has been the classical explanation of how rocket propulsion moves a craft forward inside and outside the atmosphere. Is there possibly more to the science of rocket thrust then classic science has been able to describe?

In the photo following, is a photo of a Rocketdyne F-1 rocket engine with nozzle. Five of these massive engines are in the first stage powered the giant Saturn V rockets that took the Apollo manned space craft to the moon. Generating some gigawatts of peak power during lift off, these engines, energy output was equal to that of the peak demand of electrical power of Great Britain (Hutchinson, 2013). That amount of energy being focused into a small area may lead us to an expanded understanding of how

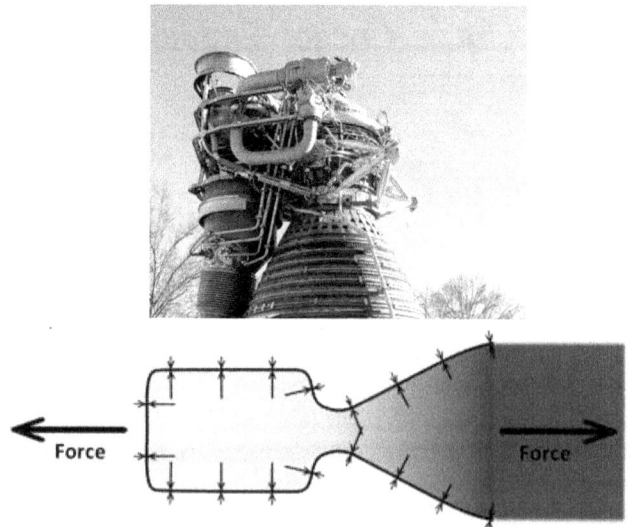

Fig. 10-7: Classic diagram of forces in an operating rocket engine

rocket propulsion works.

Early in the American rocket program, many rockets would quickly begin vacillating after takeoff and the rocket would veer off course and out of control. Watching the many movies of these early launch failures is quite fascinating and informative from an observational understanding of how rockets work. One of the most prominent features found in these pictures and movies of these early failures of the American rocket program is the loss of balance of the rocket siting on its thrust. Yet according to mainstream physics a rocket moves forward by the forces pushing on the foreword end of the combustion chamber and the exhaust is just the opposite reaction not the driving force. How then is it that a rocket can tilt back and forth on the engine nozzle, if it is actually being held up by a force pushing on the forward end of the combustion chamber? Is it possible that there is another explanation for the observable phenomenon? Is there an explanation that would apply equally to rocket propulsion systems, sails, and the simple paddles of our ancient ancestors floating on a log?

To answer this question we have to go back to the late years of the 19th Century and the early years of the 20th Century to understand the trajectory that physics took, which led us to our understanding of rocket science today. In 1887 the results of the now famous Michelson–Morley experiment was published by

Albert A. Michelson and Edward W. Morley, which apparently showed that light traveled at fixed speed and there could not be an ether that light was acting as a wave phenomenon in. Yet the experiment and the many that have followed after it with even more sensitive and precise instrumentation suffers from one very important flaw, they fail to recognize that by using light as the measuring medium they can only ever record a maximum speed of

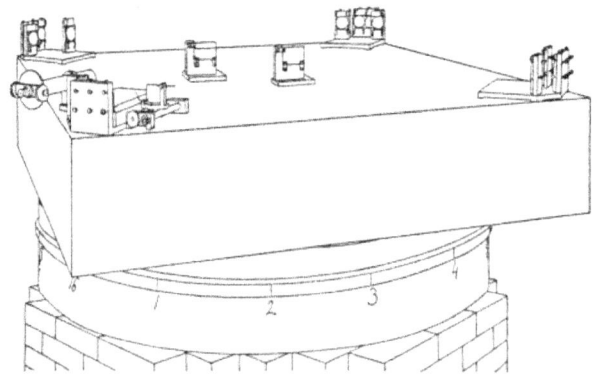

Fig. 10-9: Diagram of Michelson–Morley experiment reported in 1887 (Michelson, 1887)

that of light. We discussed this issue of faulty experimental configurations in previous chapters explaining that in order to get an accurate observable measurement of speed the medium of measurement must be traveling a speed greater than the object being measured. Because when we use light (electromagnetic

radiation) as the measurement medium for measuring the speed of light then we will always be measuring the speed/time it took light to travel from its origin to the point it was at when it hit our eye/detector. For our current discussion on propulsion systems however we need to focus on a primary conclusion of these experiments in the late 19th Century and early 20th Century and that was the conclusion that there was no either or as we describe it, an energy ocean.

In a later section of this book, we will discuss specific experiments that will be able to demonstrate the existence of energy as an actual component of our ocean universe. These experiments will demonstrate how we can use measure true speed through the universe and how we explore a new generation of space propulsion systems that take advantage of this energy ocean. So for the purposes of this chapter let us think outside the box and hypothesize that energy does exist as some type of elemental component of the universe that behave much the same way water behaves in an ocean here on earth. We like all pieces of physical matter are like chunks of ice in that ocean. With this assumption, we can offer an alternative understanding of rocket propulsion that can open the possibility to engineering systems can allow for space travel that are only in the realm of science fiction today.

When a rocket motor fires, the combustion of the fuel and oxidizer changes the molecular structure of the molecule of each. As the chemical reactions causes molecular reconfigurations of the molecules vast amounts of energy particles are rapidly released increasing the free energy density (pressure) of the localized space. This increased pressure follows a path of least resistance out the throat of the combustion chamber and into the exhaust nozzle. As the energy begins to expand outward in all directions, it creates a force against any matter that it comes in contact with. This drag force, much like that found pressing against a sea-anchor/sail, will push that matter in the direction of expansion. Another maybe more appropriate way of visualizing this force in action would be to think of a surfer riding down a wave at the beach. As the crest of the water, moves forward from behind the surfer the surfer rides upward while in a continuous path to a lower level of force. If we think of the wave as a localized area of more dense force (increase potential energy in classical physics) we can easily relate this to the plasma ball behind/beneath a rocket motors exhaust nozzle. As the energy wave moves outward from the center of the ball of the dense energy, the rocket nozzle is virtually surfing down the wave just like the surfer shown in the following picture Fig 10-10.

"Recapitulating, we may say that according to the general theory of relativity space is endowed with physical qualities; in this sense, therefore, there exists ether. According to the general theory of relativity space without ether is unthinkable; for in such space there not only would be no propagation of light, but also no possibility of existence for standards of space and time (measuring-rods and clocks), nor therefore any space-time intervals in the physical sense." (Einstein, 1920) What Einstein and others have continually failed to realize is the implication of this aspect of relativity being that "ether" or energy surrounds us like water surrounds the fish of the sea. However, this ocean of energy has characterizations that have real implications in our daily lives.

Traditional rocket propulsion theory has to be modified to accommodate the fact that energy can be forcibly localized into areas of increased density that tend to seek equilibrium with the surrounding density and as this movement of energy takes place it exerts a force in the direction of expansion on objects in its path. In the case of a rocket engine, the force comes from the release of energy from a chemical reaction between the rocket fuel and its oxidizer. The propulsion of a rocket results from the very same

Fig. 10-10: Surfer riding from state of higher force density to state of lower force density (Shooter, 2015)

phenomena that can be seen moving an object when an explosive device is set off. Objects of sizable weights can be thrown thousands of feet and even miles away from their point of origin by large blasts yet none of these objects have a combustion chamber. Just like junk floating in the ocean will be tossed around by the waves at a beach by the same waves that propel a surfer on a well-controlled ride, energy moving to equilibrium from a localized high-density area will also propel a rocket or toss objects widely. So in essence and reality, a rocket is propelled forward by the effects of a continuous explosion just behind the rocket nozzle.

Thinking back on our discussions of our ancient ancestors we have to ask ourselves did they see the effects on floating branches of throwing rocks into the water next to them. Did they see how the small waves from the rocks they threw into the water pushed these floating branches and leaves along thinking, "What if we could have a non-ending wave that could move a floating log?" and thus the idea of a paddle was born. In a way, this is exactly the process that our history of rocket science has followed. With the advent of gunpowder, people saw that effects of force/energy redistribution when they blew up large quantities of gunpowder and heavy objects were propelled through the air. It would only take a little imagination and some basic engineering to shape together the first primitive solid fuel rocket engines. Contemporary rocket engineers, like their ancient predecessors, continue to explore this basic concept with research into nuclear propulsion systems such as the Project Orion Project. This project developed under the oversight of the Defense Advanced Research Projects Agency (DARPA), which was intended to propel a spacecraft with Nuclear Pulse Propulsion (NPP) or External Pulsed Plasma Propulsion (EPPP) Technology. The concept was essentially to explode small nuclear bombs behind a spaceship and let the force of the blast push against an armored push plate at the back of the spacecraft. Velocities theoretically achievable would reduce a trip to Mars from about 3 years to a matter of a few months (Bonometti & Morton, 2000).

It is argued that the validity of the EPPP concept invalidates the classical explanation rocket propulsion as there is no combustion chamber or rocket exhaust nozzle, only a push plate, which a flattened two dimensional version of a rocket exhaust nozzle. In the following picture, Fig. 10-11, you will see three different examples of using areas of increased force/energy density to propel a vehicle along at the force/energy density tries to reach equilibrium with the surrounding universe. A surfer is pushed forward by the force density of the wave trying to reach equilibrium with the ocean surface, an EPPP rocket is propelled by the force density of the atomic bomb blasting off, and the Saturn-Five rocket was pushed forward by the force density of the burnt fuel in the nozzle looking for equilibrium. There is one difference between these three methods of using and increased area of force density to propel and object. Can you determine what that difference is?

Our surfer is using force density in a much different manner then the rockets in the other two pictures, only because unlike the surfer, the rockets are inefficiently carrying their own compressed

Fig. 10-11: Image Credits: San Diego Shooter 2015, Joe Bergeron 2006, NASA 1969

energy with them in the form of fuel while the surfer is using the ever present ambient energy. When the surfer paddles out to sea, she is merely increasing the force/energy density of small amounts of water with her hand. As the surfer rides the wave in, she is taking advantage of the increased force/energy density in the wave that is created by gravitational and wind forces.

Conversely, our rockets both are bringing their energy that they will use to create an area of increased force/energy density with them, instead of just using the ambient energy all around them.

Imagine if you will a large ship that was propelled by a water-jet (hydrojet), which is a propulsion system that pumps large amounts of water out a ducted nozzle at the rear of the ship were the propellers are usually located. This is a similar system, as is used on a Jet Ski, only unlike a jet ski, our ship analogous of a rocket, is carrying very large tanks of water on board to pump through the propulsion system. Our surfer and jet skis use the water already around them for propulsion, so that they don't have to carry extra mass. Imagine what the outcome would be if the Jet Ski or surfer had to carry water with them to use for propulsion? Their vehicle would be quickly overcome with excess mass, just as our modern rocket systems are. If ships that crossed the oceans had to carry the water they need to power their hydrojet propulsion system, they would only be able to carry a small fraction of the cargo they carry today.

In an ocean universe, we are suspended within a sea of energy, just as a boat in water or a lighter-than-air craft is floating in the atmosphere. When we bring energy (compressed in the molecules of rocket fuel of any type) with us in order to be expelled out the back of our space ship we are operating at the most inefficient method possible. Yet without an accurate scientific model of the universe, we are like our ancient ancestors that could feel the wind but had no understanding of air and air molecules. In order to be able to transit the vast distances of outer space we will have to recognize the existence of energy as a real component of the universe, just as water molecules make up the ocean and air molecules make up the atmosphere. Once we mature scientifically enough to have this fundamental understanding of energy and its presence, then we can begin to engineer spacecraft that propel themselves, not by jettisoning energy that is brought on board but by pushing against the ambient energy all around us. We need to develop for energy the equivalent of propellers and pumps similar to those we use in water and air only capable of moving energy instead of matter.

Chapter 11 – Energy Efficient Propulsion

If we are going to go "Hang Ten" with the "Great Kahuna" and we don't want to carry a wave's worth of water on our back and board, then we simply go to the beach. If we are wanting to travel to a nearby star system and don't have hundreds of years to live or a ship that can carry a near infinite about of rocket fuel and we don't want to be setting off megatons of nuclear bombs behind us, how would we propel our ship?

If we make the argument that the universe is filled with energy, like the ocean is filled with water, or like the atmosphere is filled with air, we should be able to propel ourselves through it like a boat on the ocean. It is interesting to note that our ancient ancestors invented the paddle, the sail, and the screw propeller all without the use of calculus and other higher order mathematics. With that in mind, let us examine what know about our ocean universe to determine how we could propel ourselves through it.

As we discussed earlier, if we make the argument that energy exists as actual infinitesimally small units, such that each unit of energy is many times smaller than an atom of matter. Energy units or particles (used loosely herein without inferring a matter like existence) are so small that they move freely through the densest matter. We also know in this energy universe matter displaces energy much like ice would displace water as it floats along. We discussed that the relationship between energy and matter are very much analogous to the relationship between water and ice, being the same thing in different forms. We have also established a basis for our energy displacement drive. We will need a propulsion system that does not require us to bring along energy to expel out behind us, in some fashion, in order to create a localized region of increased force/energy density pushing us along. So let us look at some potential mechanisms of propulsion for use in the ocean universe.

Many of us as youngsters, who were interested in science and physics, have explored the marvelous and mysteries phenomena associated with a spinning top or gyroscope. What makes this marvel of childhood and heart of advanced guidance system work? Can we conduct an operational analysis of this device to understand the mechanics by how it works, in light of our understanding of the ocean universe?

As a piece of matter moves through any point in the universe it displace energy, just as a submarine moving through the ocean depths continually displaces water. As the matter moves past a given point the ambient pressure of energy in the universe forces energy back into the brief void that was created by the passing of the matter. Remember energy units/particles are so small, they can pass through and permeate all the space between molecules and are trapped within the space between electrons and protons in an atom. Unlike a submarine, there is not as much displacement as the outer dimensions of the material might suggest. At slow speeds, the mass of a moving piece of matter and its apparent mass (moving displacement) is virtually the same. However, as the velocity of the matter increases, it tends to displace a few more energy units/particles and it takes slightly longer for these particles to refill the void. If we could move the matter fast enough, its mass (measure of displacement of energy) would begin to increase significantly and at some point the space occupied by the parameter of the matter would be free of all free energy units/particles (those not trapped in atoms).

What if we were to move the block of matter in circular path such as that of a rotor on a gyroscope? Now a as we increase the speed of this rotor to increasingly more revolutions per minute it will begin to displace more and more energy. As we reach a high enough speed of rotation, the individual molecules of matter in our rotor will not only displace the energy but they will being moving fast enough to prevent the energy from coming back into the area inscribed by the dimensions of the rotor. As the energy is displaced by the spinning rotor, the apparent mass of the rotor, as measured by the displaced water, increases proportional to the completeness of the displacement of the water. The following diagram (Fig. 11-1) shows a rotor spinning at a slower speed and one spinning at very high speed with donut like field of increased force/energy density. A similar effect can be seen by spinning a rotor like shape made of porous material in water.

Because of the centrifugal effect, the displaced energy is mostly discharged from the inscribed space through the edge of the rotor creating a donut like effect in the energy. As the pressure in the donut like pressure ring starts to increase, the ring structure begins to fatten, eventually increasing the density along the two flat sides of the rotor. The faster the rotor spins the higher the

Fig. 11-1: Example of energy displacement by normal rotor and a rotor spinning at ultra-high speed (Straitt, 2015)

force/energy density of this region becomes. When the density reaches a significant level, it creates a barrier to matter easily passing through it. It is at this point the device begins to experience the phenomenon we call gyroscopic precession.

In simple terms, gyroscopic precession is the resistance of a spinning body to angular changes along its plane of rotation. In figure 11-2, we can see a spinning wheel being supported on its axis at one end by a cable while the other end is free to move in any direction. Then as long as the wheel maintains sufficient RPMs, it will overcome the force of gravity to pull the free end of the axle downward. If a force is applied to the axle in any other direction, the wheel will tend to resist changing its direction of rotation. Up until a certain point, a point when a force is asserted on the free end of the axle at a right angle to the direction of

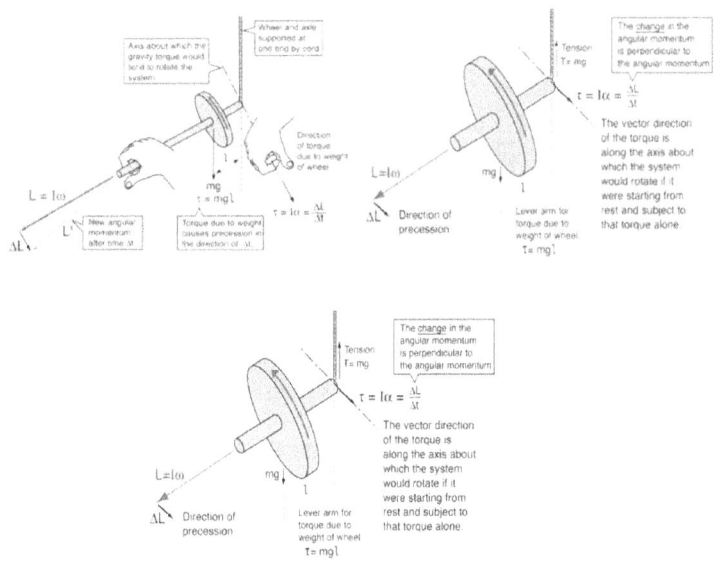

Fig. 11-2: Gyroscopic Precession (Nave, 2000)

rotation the axle will actually push back with a slightly greater force. As the outside pressure increases, eventually, the force of precession will be overcome and a new direction of rotation will be established

One of the most interesting observable phenomena of precession is easy to demonstrate with an electric motor and attached flywheel attached. If the spinning motor is suspended by a cable at one so that the flywheel is on the opposite end hanging down and the cable is twisted in one direction the flywheel will pull downward, while if the cable is twisted in the opposite direction then the flywheel will lift upwards. As long as the cable is being twisted, the flywheel will continue to climb until it actually will come in contact with the cable. By controlling the rate of twist of the cable the flywheel can be maintained at any desired elevation.

While this phenomenon can be mathematically described using Vector Rotation and Angular Momentum Formulas so as to determine the resulting force on the axle when a specific force is applied to the axel, the underlying mechanics has not previously been described. As the wheel is spinning (see figure 11-3) an area of increased force/energy builds up around the wheel exerting a force against the two sides of the wheel. As an angular force is applied to the gyroscopes axle the wheel starts to push up against this area of denser force/energy and it resists this movement into its area by the wheel.

If enough force can be exerted against the area of denser force/energy by the wheel surface, the entire donut of increased density/energy will shift to the new plain of rotation. If the wheel cannot overcome the denser area of force/energy then it will be repelled back towards its original plain of rotation. The thicker

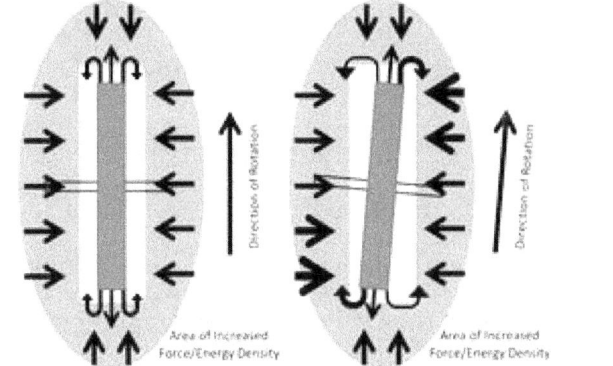

Fig. 11-3: Gyroscope normal rotation (left) Precession effect (right)
(Straitt, 1980-2015)

arrows show how increased energy flow form the edge of the
wheel to the density zone causes an increase in force on opposite
sides of the opposing edges of the spinning wheel.

The power of the gyroscopic precession is strong enough to be
useful in ground transportation. Several designs for monorail
train systems were produced and tested early in the Twentieth
Century. These monorail trains were supported on inline wheels
riding on a single track and were balanced by means of gyroscopes
embedded in the train car. By adjusting the orientation of the
gyroscopes relative to the railroad car's frame, the car could be
maintained in a level position independent of the tilt of the
railroad bed. Using the same technique the car could be tilted
when going around corners, thus providing for stabilization of the
car against centrifugal forces generated in cornering. This use of
gyroscopes for stabilizing railroad cars, as shown in figure 11-4,

demonstrates the force/energy density generation capabilities of a mass moving with high angular momentum.

What if it was possible to harness this power, to create a kind of lift, such that an apparatus might experience a phenomenon analogous to that of a sail or a wing flying through the air? A device designed such that it could take advantage of the effect of pressure differentials on opposite sides of an object. Such a device would operate in a very similar way as a propeller, in that it would move energy from one side of the object to the other, creating a moving junction of high and low pressure. This movement would cause the device to be continually falling or surfing if you will into the low-pressure region just to the forward side of the object.

In the atmosphere, the pitch of a propeller blade is used to determine which way the movement of air being moved by the propeller is directed. One way of accomplishing this would be to create a type of virtual pitch for our rapidly spinning gyroscopic wheel. If we can induce a virtual pitch for the wheel, then we can move energy from the front side (direction of lift) and create a higher density on the backside (opposite side from lift) of the wheel. The wheel will begin to behave in the same fashion as the previously discussed rotors, surfboards, and propellers. Before we go further, some readers may be interpreting this hypothesized device as a science fiction anti-gravity machine of some sort, which would not be an accurate assumption. The concept here is no more anti-gravity then that of an airplane wing or a sail on a ship. What is being described is the application of proven applied physics, used in aviation and boating every day, only now applied to the hypothesis medium of energy in our ocean universe. Two questions are to be answered. The first, does energy exist in a form that we can press against it? Moreover, the second, is it possible to create physical device that will move energy from one point to another to create an energy pressure differential on two sides of an object?

Let us look at a potential approach for creating this "energy propeller". Remembering that the matrix of the densest matter we know is to energy in similar structure as chicken wire fencing is to

Brennan's Monorail (Wikipedia, 2015)

Fig. 11-4: Scherl's Monorail Car (Wikipedia, 2015)

molecules of air. We cannot just make a simple propeller device with variable pitch props and expect to go anywhere. However, the gyroscopic precession phenomenon that we discussed may be a starting point for our design effort. We know from our observational analysis that when the wheel (rotor) of the gyroscope reaches significant angular velocities it begins to displace energy along its plane of rotation. In other words, it moves the energy from its axis outward and expels it along its outer edge. Of course, energy is trying to get back into the wheel from all points, but once there is sufficient velocity it is expelled from the volume of the wheel faster than it can reenter the space occupied by the wheel. The donut of increased energy density surrounding the gyroscope is pushing against the surface of the wheel at a force greater than the force of the ambient energy surrounding the whole apparatus. If we can decrease the energy of this energy donut on one side of the spinning rotor, the increased pressure on the other side will press against the rotor and should cause motion.

One approach that may work to move the energy incorporates the use of magnets embedded into the wheel towards the outer edge of one surface. On a frame directly above these magnets is mounted one or more electromagnets whose magnetic field can be adjusted through increasing or decreasing the electric current flow in the coils surrounding them. The electric magnetic mounted above the rotor surface will act as a transition path for the denser energy in the donut area to be focused into the spinning rotor through the magnets mounted in the rotor. Depending on which way the polarity is set to in the electromagnet there will be either a repulsive reaction or attractive reaction created between the electromagnetic fields in the permanent magnetic fields in the rotor. This will create a virtual conduit for energy flow, which will allow the energy density to dissipate from one side of the rotor faster than on the opposite side, creating a low-density pressure zone resulting in a lift effect in the direction of less dense energy. In a sense, we are creating the same type of increased force/energy density we find in a rocket exhaust nozzle only under controlled conditions.

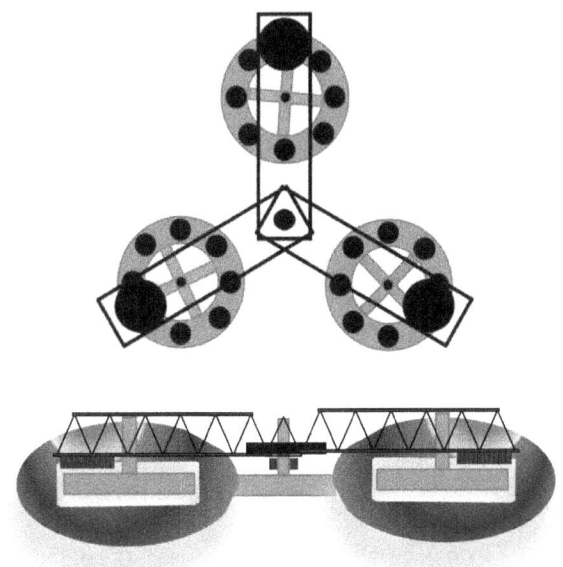

Fig. 11-5: Example of how energy lift could be generated
(Straitt, 1980-2015)

Because the electromagnetic is acting to create a conduit for ambient and locally displaced energy to flow through and the rotor is only displacing energy, neither is the source of the energy for propulsion. Because we are using the ambient energy for propulsion, the lifting vehicle only needs to carry energy for control, rotor rotation, and internal power, with none being expelled for propulsion as with a rocket. The concept of this device is to decrease the force above the spinning rotor and increase it below the spinning rotor by changing the localized energy density. This would provide a lifting reaction, similar in concept to aerodynamic lift that would move the rotor in the direction of less force/energy density.

In figure 11-5, we can see a representation of potential lifting device that has been activated. The arrows represent the direction of the anticipated forces that will be present when the devise is operating. The solid arrows indicate the direction of higher force/energy density moving against the spinning rotor from beneath. The cone or depression above the magnet represents a region of lower force/energy density above the rotor at that point. The hollow arrow indicates the direction the rotor would follow as the lifting acts on the support arm. As all three arms experience the same lifting actions leveraged against the point where the arms join in the center of the devise the net direction of lift will be in the direction of travel of the hollow dashed arrows.

At the time of this writing, the author has yet to build and test a model of this device although significant opportunities to explore gyroscopic precession have been taken advantage of. With that experience in mind, let us look at another form of the above device.

The single rotor-lifting device shown in figure 11-6 is a more advanced form of the three-rotor device previously discussed. This machine utilizes a single rotor that is suspended within a

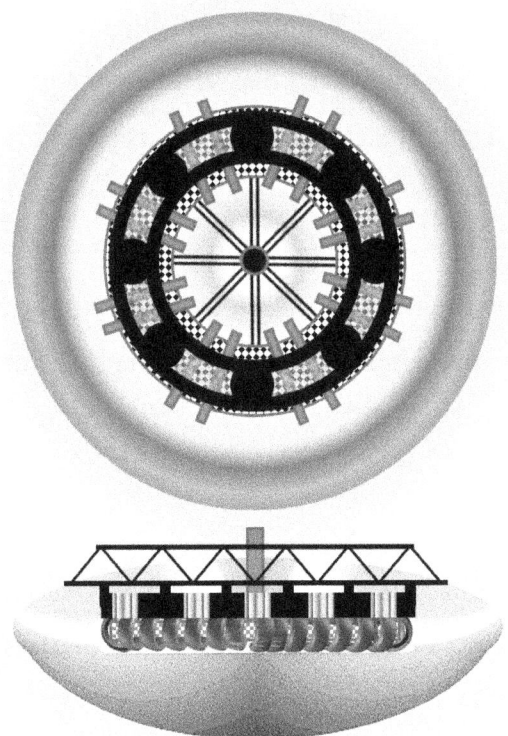

Fig. 11-6: Single Rotor lifting device with variable lift/direction control and regions of increased force/energy density.
(Straitt 1980-2015)

magnetic field that also serves as stator to power the rotors spin. The coils suspending the rotor are attached to a ring that also houses multiple electromagnets that are individually activated to provide lift and directional control. As in figure 11-5, we can clearly see the representation of the area of high-density force/energy surrounding the rotating ring. Because the figure is representative of the device in lifting mode the area above the rotor experiences lower density then area beneath it.

Just to clarify a phrase that we have been consistently using "force/energy" density. Although in the context of an ocean universe, literally filled with energy, the concepts of force and energy are not truly interchangeable, we use them in this way as a transitional phrase between the old and new physics perspectives. Traditionally in classical physics, force and energy are thought of in different ways then we define them here, with energy being the measure of the potential of work or the actual amount of work being performed. Force is thought of as a vector quantity in that it has a magnitude and direction of action. Because the ocean universe defines energy as an actual element of the universe that has properties and interacts with itself and physical matter it asserts a pressure or force against itself and matter. Thus in the ocean universe we define force as both the resistance of energy against being displaced by matter as in gravitational force, and the resistance of energy to being compressed by matter and/or energy as in electromagnet or nuclear forces. For ease of reading to those who are newly initiated to the concept of an energy filled ocean universe, in the context of our discussions in this chapter we use the term "force/energy" interchangeably with energy where applicable. If a region of space experiences an increase in its energy density then that energy, being under pressure will exert a force on the ambient energy and any matter bordering the region as it tries to spread out and normalize itself to the same pressure/density of the ambient energy surrounding it.

In the picture of the activated (figure 11-6a) device, we see that the force being exerted by the region of increased density of energy is pushing the device into the area of lesser energy pressure as energy is channeled through the electro magnets into the spinning rotor. With this configuration, a single rotor spinning at an extremely higher RPM due to being held in place by the electromagnetic field can provide more lift and better control. The next phase of development if this technology is successful would be to have a metal bullet, possibly of aluminum spinning at such an extremely high rpm within the electromagnetic field that it takes on the properties of a very dense solid rotor ring. The advantage to this is the increased ratio of apparent mass to actual mass that would be created. Ultimately, it should be possible to create a rotor out of energy or more accurately the-lack-thereof with it contained within a rotating field; this would give an apparent mass to actual mass ratio approaching the order of a black hole.

Actual mass for the purposes of the principals outlined in this theory, of an energy ocean universe, is the measure of the displacement of energy by matter or energy at rest. When matter displaces energy, which begins to happen as soon as matter takes physical form, it forces the local energy adjacent to it, outward against the ambient of energy of surrounding space. Thus, a measure of mass in a sense is a measure of energy. Apparent mass is the measure of the displacement of energy by matter because of its existence (actual mass) plus, the measure of additional energy being displaced by the matter because of its movement though the energy ocean. Therefore, an object traveling at relativistic speeds appears to be gaining mass. Apparent mass is not actual mass however and the effects of this mass are localized and limited to the time when the object is actually moving relative to the energy in the ocean universe.

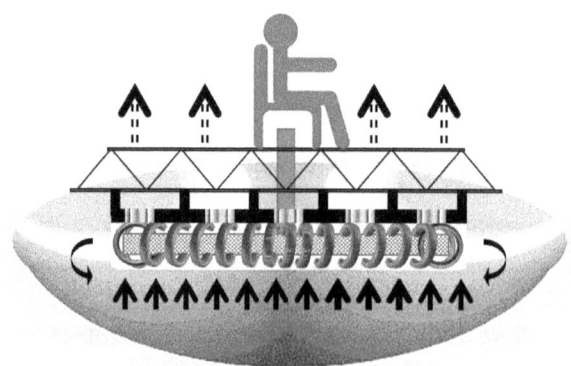

Fig. 11-6a: Single Rotor lifting device with variable lift/direction control, activated and showing direction of force and lift.
(Straitt 1980-2015)

What our lifting device uses is the creation of apparent mass, localized compression of energy, and the manipulation of the localized densities of this compressed energy to create a lift force, similar to aerodynamic lift, due to the force/energy pressure differentials on either side of the spinning rotor. As we discussed in previous chapters, within our energy ocean theory we define gravity as the force of two pieces of mater being pushed together as a counter action to displacement. In normal space, the force pushing two objects together is defined by the ambient pressure of the energy in the universe. When we spin matter fast enough to create an increased force/energy region around the object we begin to remove the object from the universal gravity and subject it to a localized gravity fields. Any object within the area of increased density will be pushed towards the center of the region because at the center of the region there will be virtually no energy.

What this means for the passengers riding on our device, they will be pushed towards the spinning rotor by the dense energy field generated around it. As the field increases to intensities sufficient to overcome gravitational forces being exerted by the ambient energy of the universe, our passengers will experience down as being towards the rotor regardless what the devices orientation is to any other gravitational field. As the passengers riding on objects on our lifting device are stationary relative to the region of increased force/energy density they will not be subject to any significant changes in inertia regardless of any change in the velocity of the lifting device and its localized force/energy region relative to any other object or the ambient energy of the universe. One of the useful byproducts of creating an apparent mass is the simultaneous creation of an artificial gravitational field/region within the area of increased force/energy density.

Another phenomenon that takes place, happens on the outer edge of the region of increased force/energy will be quite useful in space travel as well. When a boat sits in the water it displaces that water but the displaced water is able equalize its level with the surrounding water quite easily.   As the boat begins to move through the water, the water directly displaced by the bow of the boat and the water that is being displaced by the directly displaced water cannot spread out fast enough to prevent a buildup of water. This buildup of water is called a bow wave.   A bow wave has a higher force density then the surrounding water surface and anything coming in contact with the bow wave will tend to be repelled or in a sense surfed down the wave as the wave moves forward.   Looking at the picture of a ship and a bow wave (Fig. 11-7) one can easily imagine how a small boat or piece floating debris will repelled long before it could strike the ship's bow.

Fig. 11-7: Bow Wave (Alfvanbeem, 2015)

Even though some areas along the outer perimeter of the region of increased force/energy density may have considerably less dense energy than others, all of the spherical region around the device will have a significantly higher density then the ambient energy surrounding it. As particles of matter approach the lifting device they will come into contact with the denser force/energy region and just like with the bow wave of a ship, these particles will be pushed aside, surfing down the outer front of the region as it moves forward just as a surfer gliding down a wave. One of the biggest concerns of interplanetary travel, meteor and micro-meteor impacts will have been mitigated.

The Ocean Universe

Chapter 12 – Dark Lines in Spectrums

Many of us have seen the

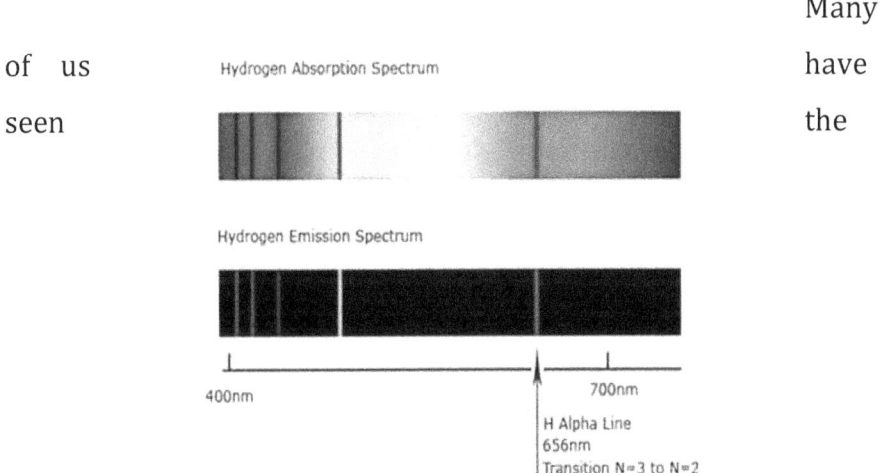

Fig. 12-1: Hydrogen Spectrum showing Absorption Lines
(Rambelya, 2010)

rainbow colored lines that appear when you shine sun light through a crystal, which is cut in a manner to create a prism effect. Fewer of us may have had the opportunity to see an actual absorption spectrum projected from a spectrum analyzer or spectrograph. Fewer still may have any idea on how these devices work or what allows for them to work at the most basic atomic and energy levels.

Starting at the macro level let us look at this phenomenon more closely. When we shine a white light, which is a light composed of all wavelengths of the visible spectrum, through a material and then examine the light with a spectrograph we are able to see the frequencies allowed to pass through the material and what frequencies of light are restricted from passing. The result is a rainbow like effect with very specifically spaced dark lines, called absorption lines, which are unique for every known element. Thus if two elements exist in a given amount of material the dark line unique to both will show up in the resulting spectrogram when a light passing through the material is analyzed. What creates these dark lines absorption lines to occur?

If an element is heated to a temperature so that it becomes illumining it admits a completely inverse spectrum. That is where the absorption lines appeared when the material has a light shining on it or through it, bright lines now appear called emission lines and the rest of the spectrum is dark. What creates these bright emission lines?

Classic physics has held that these lines occur when energy at certain frequencies are absorbed by the atom and the energy is used to move an electron from orbital level to a higher level for absorption or to a lower level for emission as energy is released from the atom. In the context of our understanding of the energy ocean universe, there may be a more mechanically accurate explanation for this phenomenon.

Popular understanding of the absorption and emission line phenomena is predicated by the belief in physics that energy does not exist as an actual element or particle within the universe. As we discussed earlier this belief is based on flawed experiments trying to determine the speed of light, which concluded that this speed is always the same relative to any observer moving with any vector relative to the light source. Yet the general understanding of most contemporary physicists is that when light passes through a substance, photons are absorbed by electrons and their energy state is raised to a higher energy of excitement.

In figure 12-2, we can see how the current theory of absorption, describes a wave of light approaching an atom. When it reaches the parameter of the atom, a portion of one wave begins to act as a particle, called a photon, and is absorbed by one of the electrons thus raising energy of the electron to a higher orbital energy state. All the rest of the light energy being carried by the

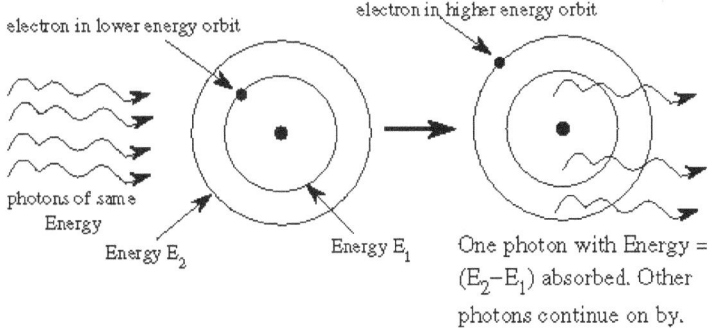

Fig. 12-2: Classical Physics diagram explaining the concept of Light Absorption (Skrutskie, 2005)

light waves continues through the atom. One of the questions that have to be asked is, "If only one photon is absorbed then why is it that the dark line associated with that specific photon and electron continues to be present in the absorption spectrum as multitudes of new photons pass through the atom?"

Is there an explanation for these dark lines in absorption spectrum that is consistent with the proposed observational model of the ocean universe? Spectrum dark lines and light diffusion consistent with the color separation the visible light spectrum is actually a fundamental necessity in the ocean universe model. In order to have a viable explanation of absorption we must also address the phenomenon of emission lines, which are the exact opposite of absorption lines. Being that in an emission spectrum, the dark lines of the absorption spectrum are now illuminated in the missing color, while the colored areas are now blackened. Of course, classical physics explains this bright line as energy being reemitted by the electron as it collapses back towards the proton center of the atom from the higher energy excitation state.

To explain these lines in the context of the ocean universe we must first understand the mechanics of how an atom exists. Generally, in classical physics, we can speak of an atom in terms of weak and strong forces, but what if there was only one force and that force was the big gravitational push acting at the micro scale.

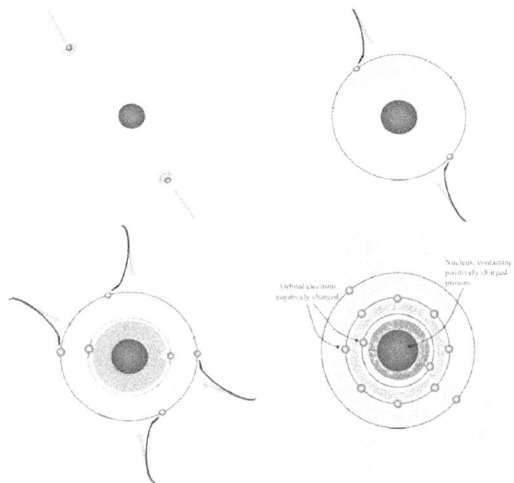

Fig. 12-3: Formation of an atom (Straitt, 1980-2015)

As we discussed earlier any two objects in the universe will be pushed together by displacement forces. As the forces, acting on the shortest distance between any two objects is always less than the sum of the forces from all other directions, these objects will be pressed together. As the smaller but rapidly moving electron with less displacement/mass comes closer to the group of protons (the nucleus) it gains more speed and its apparent mass increases thus it creates a bow-wave action, as we previously discussed, which eventually pushes up against the nucleus forcing the electron to changes it vector into a circular orbit.

In figure 12-3, we see a diagram of how an atom begins to form. The electrons are pushed towards the nucleus of protons via the "gravitational push" or reaction to the forces of energy displacement as previously discussed. This is commonly associated with the electromagnetic phenomenon of opposite charges attracting each other. As the electron approaches closer to the nucleus its speed increases and a pronounced bow wave like effect takes place (upper left figure). This increased pressure (zone of increased energy density) causes the electron to deflect and fall into a circular orbit around the nucleus (upper right figure). The gravitational forces or "gravitational push" continues to draw the electron into a tighter obit with increased orbital velocity until a point is reached where the resistance of the compressed energy within the obit of the electron balances the forces of the "gravitational push". At this point, we find that the orbital velocity (a spherical 3 dimensional orbit) of the electron is such that it can create in essence a solid barrier to the compressed energy. The electron literally is moving so fast that it reoccupies every point in the sphere faster than the compressed energy can decompress beyond the electron's orbit (lower left figure). As the sphere of compressed energy around the nucleus increase in density and density, the apparent mass of the nucleus increases as well and the newly formed atom creates an even larger region of increased force/energy density around it. This is a similar effect as we have discussed relative to a fast spinning gyroscopic rotor. As other electrons are attracted to the atom the process repeats itself until the number of orbital spheres is such that new electrons cannot be added and/or maintained in an orbital sphere (lower right figure).

Of course, our analysis would not be complete without a discussion of how the nucleus is held together with all the charges in the nucleus being positive or neutral. In the following illustration of a Helium atom Fig. 12-4, by Jorge Stolfi, what we see is the similar force/energy density area that we described both in our discussions of the gyroscopic rotor previously presented in chapter 11, and the electron sphere presented in Fig. 12-3. The small blow up frame shows the protons closely bundled together in the center of a sphere that is bound by the probability state of the electrons orbiting it. What we see in this illustration is that the particles in the nucleus are actually in direct contact with each other, unlike the electrons, which are orbiting at a relatively considerable distance from the nucleus and are not coming in direct contact with each other.

This direct contact provides the answer to the nature of the Nuclear Strong Force that is thought to hold the nucleus of same charged particles together. As we know like charges repel each other while opposites attract. In our earlier discussion of gravity, we discussed the analogy of a ball floating on water versus a non-wetted ball floating on the surface of a body of water. The concept discussed identifies the nature of displacement forces acting on the displacement causing body. As the bodies become closer together, the displacement forces increase. When the bodies are at their closest point, which is when they are touching each other, then the maximum possible force holding them together is accomplished.

Once two protons or neutrons are formed or collide with such force as to overcome their like charge and the bow-wave like force/energy density region formed by their movement then they

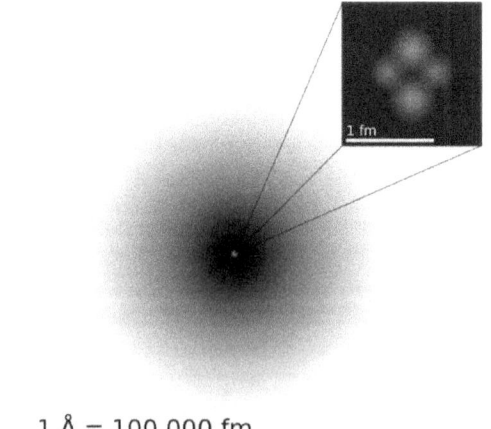

1 Å = 100,000 fm

Fig. 12-4: A representation of a Helium Atom (Stolfi, 2011)

will be held together by the force of displacement of the whole universe pressing against them. So there are the Strong Nuclear Force and gravity one in the same thing? The answer to that is essentially yes they are the same thing, only on a much different scale. When two large multi-atom/molecule sized objects are touching each other, there is an immense volume of space between the particles of matter comprising them. This immense volume is filled with energy particles that are much smaller than the atoms comprising the matter in question. Thus, the displacement of energy by these two objects is not a total displacement of all energy within the volume objects, and the gravitational force is proportional to their masses. When two protons are pressed together in this manner, they virtually have the weight of the whole universe pressing down on them. Unlike holding two bricks together where there is a volume of space between the molecules, the space between two protons is virtually nonexistent. Without any space between these two pieces of matter, there is no energy particles separating them and offsetting the displacement forces from all other directions.

Protons weigh in at some 1836 times the mass of an electron, while the proton is thought to have a diameter of about 1.6 to 1.7×10^{-15} meters, the best that can be said about the size of an electron is that it has a diameter of less than 1×10^{-13} (Pauling, 1964). While we generally think of an electron as being smaller in size then a proton, smaller size would equate to smaller mass, there is an alternative view on the size of an electron. According to Malcolm H. MacGregor, *"The electron is a point-like particle-that is, a particle with no measurable dimensions, at least within the limitations of present-day instrumentation. However, a rather compelling case can be made for an opposing viewpoint: namely, that the electron is in fact a large particle which contains an embedded point-like charge."* (MacGregor, 1992). If MacGregor is correct, and the electron is a large particle like a proton, then why would there be such a difference in mass between a proton and an electron?

We explored the concept of apparent mass in our earlier discussions of gyroscopes and learned that as the density of energy increases then the apparent mass of an object displacing that more dense energy increases, because the amount of displaced energy is more in a region of increased force/energy density than in the ambient energy density of the universe. As the electron compresses the energy within its spherical orbit, the actual amount of the energy displaced by the enclosed proton increases, and thus the proton apparent mass increases proportionally. If our theory of energy ocean universe holds true, then we can postulate that the strong nuclear force that holds the nucleus of the atom together is the same gravitational pushing force that holds the solar systems and galaxies together. Only when we are working on this subatomic scale the relationship between the gravitational forces to the size of the matter particles is much stronger.

Getting back to the main topic on the dark lines in absorption spectrums we must also include the corresponding bright lines in the emission spectrums which create an exact inverse picture of the absorption spectrum. Either both of these lines have to be created by a single continuous blocking (absorption lines) /emitting (emission lines) events or a string of repetitive events such that it appears to be a single continues event.

In the case of the emission lines, we can see in figure 12-5, that classical physics the lines are generated by the collapsing of the electron from a higher energy state to a lower one as shown the emission diagram from Skrutskie. One problem with this explanation of emission is how the electron goes from the emitted state back to the higher state of excitation during the emission

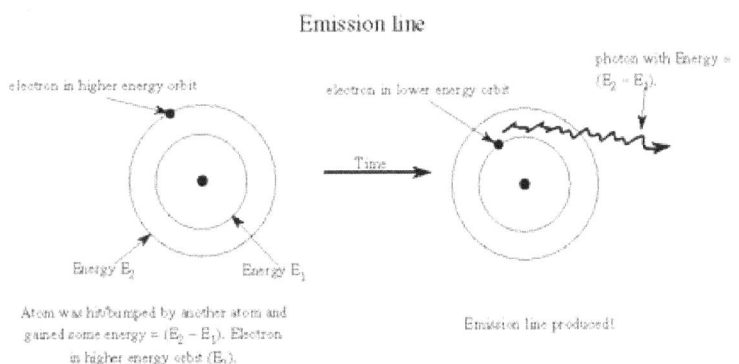

Fig. 12-5: Classical physics diagram explaining the concept of Light Emission (Skrutskie, 2005)

process, as this then is absorption. Can we feel comfortable that both absorption and emission can take place at the same time and produce an emission spectrum over an absorption spectrum?

One argument is that for a body to be illuminative it must be subject to increased energy levels such as heat or electromagnetic radiations. If this is the case, are we to believe that the electron is yo-yoing back in forth between states at a frequency of infinite magnitude, such that it appears in both states simultaneously? Perhaps not, because there is another explanation for these lines that operates consistently with the hypothesis of the ocean universe.

In figure 12-6, we see an illustration of an atom with several sine wave lines representing electromagnetic waves approaching the atom. The diagram illustrates how some of the waves are able to transit through and/or around the atom without being stopped while other waves are not able to transition the area occupied by the atom because their frequency corresponds to that of the

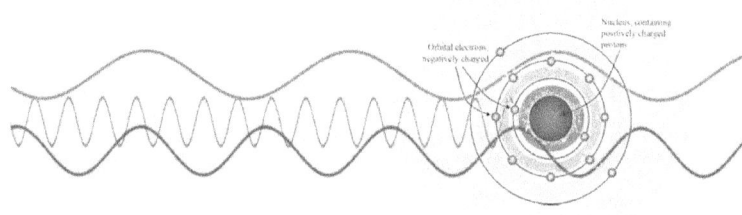

Fig. 12-6: Light waves moving through an atom (Straitt, 1980-2015)

orbital frequency and diameter of electrons. In its simplest form, this is how the mechanics of light transitioning through an atom would look. Yet, there are still more interactions that must be accounted for.

Obviously light going through a piece of matter observable by a human eye, will come into contact with many molecules. A condition that would be too complex to diagram here in its totality, but one whose effect can be extrapolated from the simple diagram of a single atom. As the light, at frequencies sufficient to transition the energy shells, begins to go through the atom it encounters areas of more dense energy, which speeds up the velocity of light, and then as it exits, it returns to its normal speed relative to the ambient energy density. This light results in different waves of light of the same frequency coming out of the atom with different timing than the same frequency that did not encounter the atom. Thus, these two waves become phase shifted or out-of-sync with each other. Another effect that takes place within the atom is refraction of the light waves as they transition the boundary zone between the lower and the higher density energy zones.

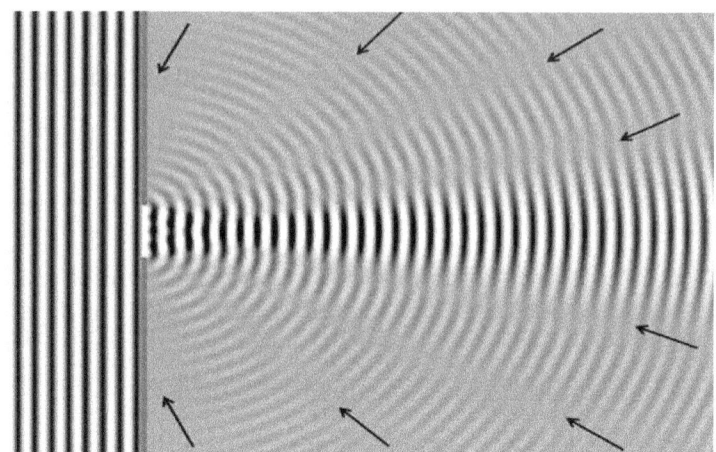

Fig. 12-7: An example of diffraction (dark spectrum line equivalents) in water waves passing through a transition point (Wikimedia, 2014)

In figure 12-7, we see an analogous representation based on a water wave being passed through a small opening in a barrier. Notice how the diffraction of the wave creates an interference pattern of lines. There is no absorption of water by the wall and there is no moving the wall to a higher level. What is happening in figure 12-7, is a real interruption of the wave action by an impenetrable barrier.

If we move to more of a macro scale, we can understand that this process also takes place when we deal with matter on a large scale dealing with thousand, millions, and billions of atoms. What we see is that one atom by itself and millions of similar atoms together produce the same absorption and emission patterns. This happens for two primary reasons. First, all of the atoms in clusters of matter of the same element are operating identically to block certain frequencies of light and allow other frequencies of light to pass. The space between atoms in a cluster is so vast in comparison to the size of the individual energy particles that occupy this space that it allows each atom to act on the light individually and separately from other atoms. What does happen though, because there are so many atoms, then any beam of light shown on a piece of matter will have all portions of that beam of light come into contact with at least one of these atoms.

For over a hundred years now, scientists have been trying to explain normal wave actions and interactions of light with highly complex and convoluted theories because of the erroneous assumption that there is no ether. However when we realize that $E=MC^2$ by its very nature demands that energy exists as a viable and interactive part of the universe, it becomes apparent that the specious wave/particle theory of light must give way to a more practical and accurate theory of a wave action in a known medium. The energy we speak of in so many theories and practical

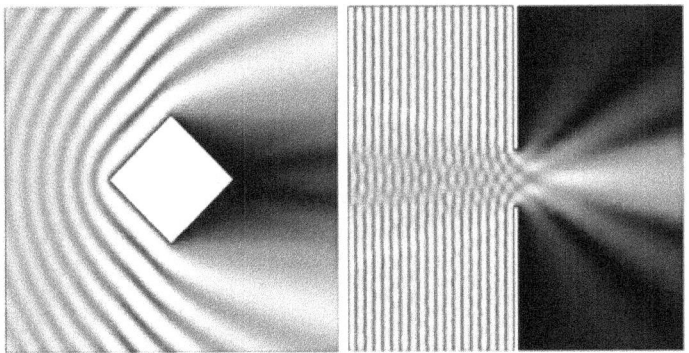

Fig. 12-8: Diffraction, an example of how light interacts with matter in an ocean universe (Frei, 2014).

applications is a tangible and elemental part of our universe, just as water is a tangible part of an ocean, or air is a tangible part of the atmosphere, allows us to transcend the previous barriers imposed on modern physics by the particle to wave to particle transformation theories. In The Ocean Universe theory, we have now reduced the complexities of the theoretical explanations of electromagnetic wave interactions with matter and other electromagnetic to a more simplified and well-known classical wave in a medium model.

Before we leave this chapter we need to discuss in a little more detail the mechanics of how exactly how the absorption and emission process works at the atomic level. When an absorption pattern appears several things occur as we discussed, light is deflected, light is reflected, and some of the energy particles may find themselves compressed into the orbital spheres of the electrons. Most of these actions can easily be observed by watching wave actions on the surface of a body of water. If we understand that energy is a medium that exist in the universe, just as water exist in the ocean, or air in the atmosphere then the same wave dynamics we see in our examination of waves in these mediums can be applied to light. How is it that we can explain the emission patterns?

When we apply energy to an atom, we are doing two things that change its dynamics. First, we are increasing the ambient energy density eccentrically in both space and time. That is as we force energy into the 4th dimensional space occupied by the atom the action is not completely uniform. The energy comes from one or more vectors but not from all possible vectors evenly whether we use electromagnetic waves, thermal, or impact forms of force it happens in some sort of pulsed format. Second, when we force additional energy into an atom, we are swelling it much like blowing up a balloon, or heating a sealed container of water. At some point the containment mechanism can no longer hold the pressure and something has to give.

In essence then, an atom when in the emission state, it can be thought of as a self-contained radio transmitter and antenna combination that radiates electromagnetic radiation in various wavelengths of the spectrum, including those in the visible light portion of the spectrum. Because of this, an atom should be found to comply with the same radiation equations for power, gain, and radiation patters found for other antennas. An initial and only cursory investigation into this matter suggests that an atom has similar antenna characteristics associated with the Kraus Spherical Cage Antenna (Kraus, 1956) because of a similar instantaneous radiation pattern. An atom also simultaneously displays the characteristics of a remarkably dissimilar antenna design, which is a phased array antenna. Let us see how this is possible and these two dissimilar characteristics interplay.

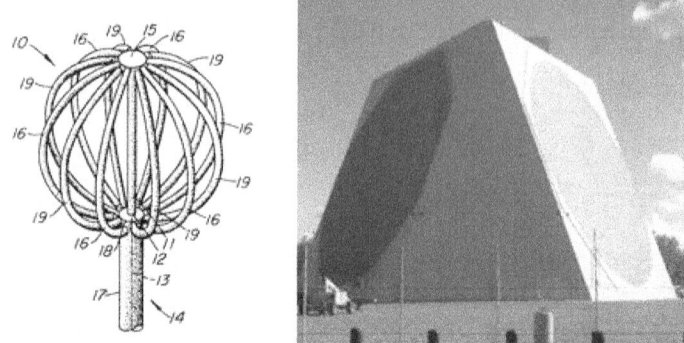

Fig. 12-9: Left, a Kraus Spherical Cage Antenna (Kraus, 1956); Right a Phased Array Antenna used for the Pave Paws Radar System (Dranidis, 2003).

A Spherical Cage Antenna has a unique property such that almost all of the energy applied to the antenna is transmitted out in a horizontal circular radiation pattern. If we were to look at cross section of the radiation pattern, it would appear as a figure 8 laying on its side with the spherical antenna at intersection of the two lobes. If we could capture an instantaneous 3-dimensional image of the radiation pattern, we would see a donut like shape of radiation surrounding the atom. If we were to move forward to a next instance and view the radiation pattern, we would see that it is still the same only it will have rotated to a new orientation. If we could follow this process though an infinite amount of steps, we would see that spherical radiation pattern develops. Within this radiation pattern would be found all of the characteristics found in any instantaneous snapshot of emission patterns.

In figure 12-10, we see a diagram a diagram from Kraus, representing a cross section of a radiation pattern from a spherical antenna showing its tipped figure 8 structures. Extending this 2-demensional cross section into to its 3-demensial volume, we see a donut structure surrounding the spherical antenna. This representation is shown horizontal, as the antenna would be using the earth as a ground plane. We then have a representation of how the spherical antenna pattern would be shifting in 4-demensions as the electrons orbit around the nucleus shifting the natural plane of radiation of the atom. Ultimately, the effective radiation pattern becomes a spherical radiation pattern.

To understand better how the moving electron can create a 4-demensional radiation pattern, we can investigate the concept of a "Phased Array Antenna". The optimum size for an antenna is not necessarily a length equal to the length of the wavelength of the frequency being transmitted. More often than not, the optimum length of an antenna is one quarter of the wavelength of the frequency being transmitted. This type of antenna, called a quarter-wave antenna, is often preferable because of its smaller size and exceptional performance characteristics. Nevertheless, smaller is not always better. As antennas get smaller than a quarter-wave, the efficiency of the Antenna to radiate power in the form of the desired electromagnetic waves also decreases. An antenna element operating at say 1/100th of the length of the wavelength of the frequency it is radiating will be very inefficient, because some of the power is reflected back to the antenna, where it is received and conducted back to the transmitter. However, there is an antenna configuration, in which very small antenna elements are arranged in such a way that energy is not reflected back towards the antenna but is redirected in the desired direction of radiation. This antenna configuration is referred to a phased array antenna figure 12-9.

A phased array antenna is an antenna that consists of a multitude of small antenna elements, much smaller the then the optimum quarter-wavelength size. These elements are arranged in a matrix and interconnected so that the degree of power sent to each element can be independently varied. The interaction of the radiation from all these elements, allows effects of wave interference both constructive and destructive to focus the radiation patter of the antenna in the desired direction. In effect, a virtual form of diffraction is created, where instead of a the electromagnetic wave hitting and going around an object or going through a slit in an object, the waves coming from the antennal elements interact with each other, as if they are experience real diffraction.

In a passive phased array antenna, the radiation pattern is fixed by the physical layout and the electrical tuning of the antenna. In an active phased array antenna the direction and shape of radiation pattern or primary axis of the resultant beam is controlled electronically (figure 12-9). It is the characteristics of the active phased antenna, which we are interested in for our understanding of the emission spectrum. In practice, a flat phased array antenna can cover a half-spherical area in front of it, without having to physically move the antenna.

An atom acts like a physically variable phased array antenna, with the electrons acting as the antenna elements. Unlike an electronically variable antenna, an atom is a physically variable antenna because while the electron has a fixed radiation pattern, its position in the orbital sphere is continually changing. The electron can be thought of as an Omni-directional antenna with the radiation directed towards the nucleus being continually directed away from the nucleus. Thus as the electrons move at relativistic speeds in their spherical obits, they create a variable spherical phased array antenna, the resulting radiation pattern being a sphere, with a net interference pattern resulting in effective radiation being in the frequencies represented by the bright lines in the atom's emission spectrum.

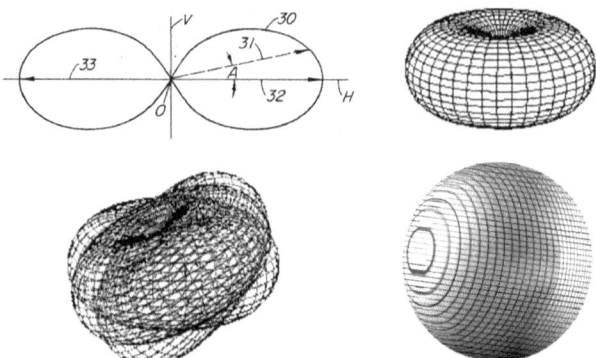

Fig. 12-10: Upper left, Vertical cross section of a spherical antenna radiations pattern (Kraus, 1956); Upper Right, A 3-dimensional view of the radiation pattern (Bevel, 2009-2011); Lower Left, Representation of several n-states in a random rotation of a radiation patter from a spherical antenna; Lower Right Representation or a spherical radiation pattern that would be created by a spherical antenna with a phased array of elements embedded in the spherical surface, or an electron distribution around an atom, randomly shifting the plane of radiation with a random pattern (Straitt, 2015).

The question is, can light act as a radio wave when it comes to antenna technology?   We find an affirmative answer to this question with what are known as nantennas.   In 1973 Robert bailey and James C. Fletcher patented the first nantenna or electromagnetic wave converter (Fletcher & Bailey, 1973), the patenting of the first nantenna was specifically described as being for converting light to electricity by Alvin M. Marks in 1984 (Marks, 1984).   A nantenna that experienced resonant light absorption and rectification in the visible range of the electromagnetic spectrum was successfully demonstrated by Guang H. Lin in 1996 (Lin, et al., 1996).   Today, nantenna technology is the being viewed as the potential future of solar energy technologies around the world. Being able directly receive electro-magnetic waves in the visible band, even the infra-red, will allow for increased power output from smaller packages.

If light can be received by a nantenna it can also be transmitted by an antenna, and this is exactly what an atom does when it is excited by electromagnetic or thermal radiation being applied to it (Maksymov, et al., 2012). In an atom's stable state, equilibrium has been established between the ambient energy environment around the atom and the ability of the electrons to contain the level of "compressed" energy within their orbital spheres. It has been established by Daniel Dregely, of the University of Stuttgart, that it is possible, feasible, and desirable to build a 3-demensional nantenna (Dregely, et al., 2011). A simple order magnitude of scaling would bring us down to the atomic scale and the application of the same principals behind a nantenna to those of absorption and emissions of an atom.

Another spectral phenomenon in support of the theory of an Ocean Universe is the mono-energy line emissions of so-called "dark matter". Dark matter is theorized by some physicists to make up, the difference between the observable versus unseen but calculated matter in the universe. Of course, in the ocean universe, it is not necessary to create an imaginary form of matter because the concerned variance in mass, between visible matter and calculated mass, is known to be the delta between the displacement of energy by matter and the self-displacement of the energy itself.

For example, Alexey Boyarsky and his team, from Cornell University, discovered an unexplainable mono-energy line emission at ~3.5 kiloelectron volts (keV) in the X-ray spectra of the Andromeda Galaxy and Perseus Galaxy Cluster. Although this emission line does not show up in deep space scans, its presences in Boyarsky's survey of the region of space around these galaxy clusters has so far been unexplainable (Boyarsky, et al., 2014). Further review of the data by Signe Riemer-Sorensen, of the University of Oslo, suggest that majority of the energy may be coming from known sources and that the remaining energy after these filters have been applied most probably rule out dark matter as a point origin for the lines in question (Riemer-Sorensen, 2014). According to Riemer-Sorensen there is not enough evidence in the data to support "dark-matter" being the source of the emissions. Is there an answer for these emissions in the ocean universe theory?

While this is not the proper forum to prove or disprove either the existence or source of these apparently unexplainable emission lines, we can offer a research direction based on the ocean universe model. In the figure 12-11, we see a real world example of how wave interference occurs when waves from multiple emission sources interact with each other.

Notice how not only do the original waves leave the region of interference and continue on unchanged, where the interference occurs a new set of wave patters also propagate outward. Some of these waves are formed from constructive (additive) interference, these are commonly referred to as rogue waves because they appear apparently from nowhere, and they are much larger than the waves that formed them. The graph and accompanying simulation (figure 12-11) further demonstrate how constructive interference can create otherwise unexplainable waves. While the patterns in this graphic are more demonstrative of a large rogue wave, interference can also create a wave of lesser amplitude

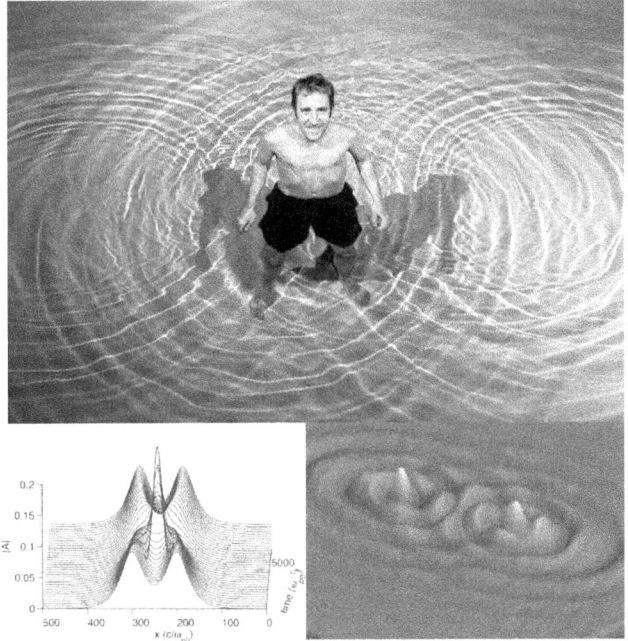

Fig. 12-11: Swimming Pool Interferometry, notice the rogue waves, (Alexander, 2014); Graphical Representation of a constructive inference with additive amplitude (Veldes, et al., 2013); Simulation of constructive inference resulting in a rogue wave (Janssen, 2009).

(energy) then either of the original waves, and this is destructive interference.

Astronomers do not spend all their time in pools studying wave interactions however; they are also scanning the skies for scientific data that will help us to understand better the universe we are a part of. As a result of this research, astronomers at NASA

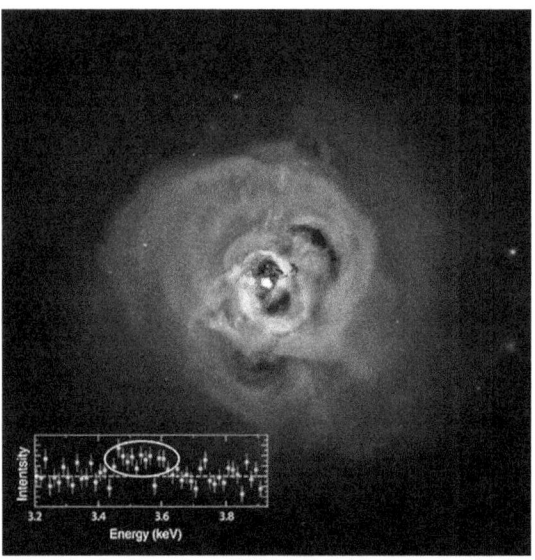

Fig. 12-12: Chandra+ XMM full field (Bulbul, 2014).

have uncovered over 70 galaxy clusters with similar x-ray signatures that display the abnormality spike at about 3.5 keV (Bulbul, 2014). While researchers trying to validate the concept of "Dark Matter" are prone to point to this data as evidence of the existence of this unseen theorized component of the universe, perhaps there is a more simple answer?

Riemer-Srensens conclude that the x-ray radiation of the Milky-Way galaxy does not contain significant enough intensity at the 3.5 keV to support the apparent spike, unless multiple observations are overlaid or stacked on each other or interfered. Interference wave observations are what the very nature of the Chandra and XMMM-Newton observational projects are about. By looking at the combined/interfered emissions of a broad base of x-ray sources, rather than selective observations of single point sources, the expectation is a better fidelity of radiation patterns can ultimately be obtained. Of course, if we look at the data from a perspective unbiased by the search of universal dark-matter, there is another and possibly more obvious reason for the peaks in the energy intensity and that is constructive interference of the multiple wave sets coming from the significant x-ray sources within the galaxies. In an ocean universe model, these peak amplitudes could be expected and should be projectable, as well as destructive interference events such as the lower intensity energy levels that are being observed at 3.7, 3.85, and 3.91 keV levels. While peaks might be explainable by the presences of what is referred to as "dark-matter" emissions, the associated troughs are less amenable to explanation by dark matter. Peaks and troughs in wave amplitudes are both common characteristics of wave interference patterns caused by multiple wave fronts interacting in a medium such as that proposed in the ocean universe.

From the subatomic to the galactic, we have been able to relate the various observed phenomena of light as it interacts with matter and with itself, by applying a common set of rules and laws governing waves in a medium, which physics is already well versed on and familiar with. By recognizing that energy, while not matter is as much an equal and accountable elemental component of the ocean universe, then we are able to unify the gravitational and electromagnetic forces at both the atomic and galactic scales.

Fig. 12-13: Upper Left, Wave radiation from the Perseus Galaxy Cluster (Cosmic Astronomy, 2014); Upper Right, Wave radiation pattern from Andromeda Galaxy (Cosmic Astronomy, 2014); Bottom, Interference patters between wave fronts caused from colliding black holes (Henze, 2009)

Chapter 13 – Time

No review of physics and cosmology would be complete without a discussion on time. Time is absolutely central to our own being and our study of the universe around us. What is time, how does it occur, and how can we manage or interact with it like other elements in the universe? These are questions scientists and philosophers have spent untold hours in contemplation and debate of. In this chapter, we will explore the mechanics of time in relation to the ocean universe, trying to understand better, how time operates and how we can interact with it in this reality.

Time as our companion is our "best friend", as our adversary it is our "worst enemy", and in all cases, it is a mystery beyond our normal senses to comprehend. Time has been the component of our legal systems, our mythologies, our religions, and our most historic moments, yet for all the interaction, we always find ourselves as its slave and never its master.

In the most simplistic terms, time can be thought of as a sequencing of events that occur in the universe. This explanation of time leaves us with a lot unexplained details that have led to questions, which continue to challenge the best scientist and philosophers of our time.

One such question, for example, concerns the baseline unit of time. While we have identified a "second" as the basic unit of time, can we be sure that this unit is valid throughout the universe? Common belief is that the ancient Babylonians were the ones to have established the 60-second minute as well as the 360-day year. However, this way of keeping time may have preceded them by millennia. For now, we can only surmise, based on ancient written records, ancient calendars, and mythology as to why at one time the earth had a 360-day year, which was dividable into 12 months of 30 days each. One outcome of this ancient system is that the length of a second was established to be a dividend of this annual duration. And, while we have improved our technical ability to measure this basic unit of time, a second, the 60 second minute, and 60 minute hour have become so intrinsic in our perception of reality that all attempts to change it have failed to be adopted for any extended period of time.

From sundials and candles, to quartz and atomic clocks accurately tracking the passage of time has been always been an imperative for every civilized society. Today we have even fabricated a process for using the radiation patterns of the Caesium atom, using the change in state of the energy between two electron spin energy levels or more specifically 9,192,631,770 cycles of the radiation emissions. From a cosmological perspective, the problem with this way of defining time is that we set an arbitrary length and then went and manipulated some other natural occurrence to fit our length. Some important questions have to be asked such as if our arbitrary length of time units are first, a natural occurring measure of time, and second, does the unit have any universal validation? Another way of looking at those questions would be, "If an advanced alien race from a distant world would recognize our time keepings system as scientifically valid?"

For comparison, let us look at an analogous system, our system of temperature scales. One of the first standardized systems of measuring temperature is the Fahrenheit system of measuring temperature was proposed by a German Scientist, Daniel Gabriel Fahrenheit, during the first parts of the 18th Century. Using a sort of disjointed set of references, he concluded that the temperature at which brine froze would be zero degrees (0°) and the average core temperature, which the human body maintained, would be 100 degrees on the Fahrenheit scale. By using, an evacuated tube filled with mercury the distance between these two points on the scale was then divided into 100 equal parts. Thus, we have water freezing at 32 degrees and water boiling at 212 degrees. The Celsius scale attempted to do the same thing only using the freezing point of pure water as 0 degrees and the boiling point of water as 100 degrees. Like our unit of time the second, these measures of temperature are relatively based on physical phenomena that existed to which we try to build a measurement system to justify. However, another system exists, which is not arbitrarily defined, and is an absolute measure of temperature. Temperature by definition is an indicator of atomic/molecular motion. The more molecular motion the higher the temperature of a substance, while the less molecular motion the lower the temperature. What if there is no motion? When a there is no atomic/molecular motion the temperature of an object is nothing. Or in the Kelvin scale absolute zero, which is some 273 degree below zero on Celsius scale.

The power of this method of measuring temperature is that it is based on an absolute position rather than a relative one. For example, the boiling point of water is relative to other factors. For example, raise or lower the atmospheric pressure of water as little as 30 cm and its boiling point will change inversely by about 1.1 millikelvin degrees. This is like the preverbal task of trying to hit a moving target from a moving platform with an independently moving gun, in the fog. So using the freezing point of water to set a temperature scale and a second to set a time scale is iffy at best. Whereas using an absolute point such as absolute zero is consistent, if not readily achievable. Can our measure of time be also based on an absolute, such that there is a scale that records zero time passage and then some infinite value at the other end of the scale? If so then what is the medium of this change? With temperature, we are looking at some sort of change in the state of matter, such as air, water, rock, steel, etc.; so what would be changing that could create a change in what we call time?

Just as we can determine if something is getting hotter or colder, with any of the common temperature scales, or even your grandpa's proverbial internal hot and cold meter, so can we measure the progression of time with any standard, even if it is not perfect. The one thing we know about time, which is interesting, is that when you go from an intense gravitational field to one of less intensity, time seems to slow down for those in the less intense gravitational field. We also know that time seems to slow down for someone who is traveling at a high rate of speed. Is there a connection between these two conditions, such that a common causation is acting in both cases?

In the Ocean Universe Theory, we remember that gravity is not a pulling force but a pushing force. Gravity is the reactive force to energy displacement, just as buoyancy is the reactive force to water displacement. When we look at the forces pushing from all directions against the hull of a boat, we can calculate that the vector of the sum of these forces is greater than the opposing downward force of gravity. In the Ocean Universe Theory, energy exists as an elemental component of the universe that can be pressed against by matter. Thus, when matter exists, it displaces energy and pushes it out away from the spot matter is occupying. Of course, all the energy in the universe contributes to this reactive force, which then pushes back against the matter and thus we experience gravity.

Previously we discussed when matter displaces energy there is an area of more dense energy surrounding the matter and reducing back to the ambient universal energy density levels. When we discuss temperature, we are talking about the exchange of information from one molecule to the next through vibrations. Literally one molecule is activated to vibrate and passes on that information of vibration to the next molecule and the next etc. This process happens millions of times a second, with many starting and receiving molecules, which continually switch roles. Can there be a familiarity between the phenomenon of temperature and the phenomenon of time?

We can think of time as more than a placeholder for an event that is to occur, it is a transfer of knowledge or information. As each minute event of the universe occurs, information about that event is constantly being transferred from one globule or photon of energy to others that surround it. This transfer takes place at the speed of light but it must go from one photon to the next in a sequential pattern. It is this progression of information transfer, which establishes our concept of time. So in essence, our concept of time is analogous to the concept of temperature, in that many globules or photons of energy are passing along information about events from one to another.

If time in our ocean universe is a sort of information transfer from one photon to the next than how do we explain not only the apparent unidirectional flow and what appears to be varying flow rates that seem to systematically change under different conditions? We will see that the same characteristic of our ocean universe, which produces the observable phenomena of light, also inversely produces the consequences of time that control how we perceive the universe around us. The condition of increased density of energy particles in a region creates an increase in the speed of light waves in that region, and the same condition slows down the transition of time in that region. Let us look at how this happens.

While light waves represent a simple propagation phenomenon, time propagation happens in a different way. Time, like temperature, is a process in which each data point has to be transferred from one particle to the next in more a sequential fashion. Once a ring of particles at a given distance have received the information of time, then the information front moves on to the next ring, and the next ring, and the next ring, throughout the universe. And, while light can be thought of as an analogue wave that dissipates its force as it travels, time can be thought of a digital transfer of information. Thus, time information can be transferred across the universe without any loss of information or fortitude as it disseminates outward from the events that took place, which the time is recording.

In a region of ambient energy density, the space between energy particles or photons is such that for every cubic unit of space there are a fixed number of energy particles or photons. While time travels between particles at the speed of light it must also travel to every particle in the given volume of space. Thus the more particles of energy in a given unit volume of space the more connections that are made and the longer it takes the time event horizon to move outward from the origin point. This transfer of information along with the embedded delay is analogous to sending out a message across a multimode network. We know that in digital switched packet communications systems, delays in transfer of information from end-to-end grow as the number of nodes increases. An example of the type of delay that can be expected in a multi-node system is that of Paterto Traffic Sources. If for example, "we have a Pareto traffic source with tail index α we find that end-to-end delays grow as $O(N\alpha+1/\alpha-1 \,(logN)\,)$ with the number of nodes as $O(\frac{N\alpha+1}{\alpha-1} \,(logN)^{\frac{\frac{1}{\alpha}}{\alpha-1}}\,)$ with the number of nodes N (Liebeherr, et al., 2010).

Of course, as the number of nodes decreases the speed at which information can get through a multi-node system increases. In our ocean universe, we understand that the density of energy particles around an object of matter, or displacement action caused by an abnormality in the ambient energy matrix, such as a void bubble (black hole) behaves very similar to that of matter particles in compliance with Fick's Law, which requires that particles move from a high concentration to a lower concentration. While we will observe a higher density of energy particles in the displacement zone around an object, we will see a natural reduction in this density as we move outward towards the ambient universal density of energy. The farther we get from the "mass" generating displacement event, the faster time information moves through the region.

In essence, what we see is the reduction of the time and space problem down to a communications about events within the fabric of space problem, by effectively looking at time as a communications mechanism. Time is a process by which activity in the universe is not only recorded but it is transmitted/propagated through a multi-node network much like they synoptic network one would find in a human brain. Time as we traditionally think of it is not just a sequencing of events rather it includes recording and broadcasting of the information about those events across the universe. Where this train of thought carries us to, is an even more profound concept when we begin to understand that what we think of as time may also be a mechanism to cause actions to take place. In other words, time is already a two-way street, much like what is referred to in the controls world as a Supervisory Control and Data Acquisition System or SCADA system. A SCADA system not only monitors and records events in a system, it also initiates actions and can direct the course of those actions. Additionally, just like our understanding of time, a SCADA system moves continually forward and never backwards.

While a SCADA system may allow us to repeat and relive an activity in the system, each of these actions take place sequentially after or later than the original activity being repeated. If an activation/deactivation event is repeated, it is not that we literally go backwards in the sequencing, we actually add to the master sequencing of events a repeat performance of the prior event. We also must understand the future events also exists within the SCADA system in the context of information about the system itself and the information held within the matrix, as well as, events planned to initiate at a future point. We will get back to this concept later, so let us return for a moment to the more basic and observable element of time that we are familiar of.

Because we know the rate of time is not fixed, that is time passes at different rates, depending on different physical properties of the space it is acting within, we should be able to explain this observed phenomenon in simple form. We discussed how and why the rate at which time acts would be slower or faster in denser or less dense energy (gravitational) fields, but why would it appear that the rate of time is slower for a frame of reference moving at a higher rate of velocity and lower for a frame of reference moving at a lower rate of velocity? The answer is remarkably simple and has less to do with actual velocity then it does with the concept of apparent mass that we discussed previously in the discussions of propulsion systems.

Remember, how we showed in figure 11-7, a ship in the ocean moving forward through the water would create a bow wave? This bow wave can be thought of being analogous with what happens as a body of matter moves through the ocean universe. Just as the ship will begin to create a larger and larger bow wave as it moves faster through the water, a body of matter will create a corresponding zone increased energy density around it as it moves through the energy in the ocean universe. When we relate the height of the bow wave of a ship to force density we can then see how we would have a force density increase around our body of matter, which would result in a region of increased energy density directly around the moving body of matter. Remember this is what we previously explained as the increase in apparent mass versus actual mass. This increased energy density region is very similar to what we would see around a very large and/or dense body of matter such as the earth or sun, and as with the region around the earth, the increased density zone would be traveling with the moving body. In fact, this happens anytime a body moves through the ocean universe, only at relatively slow speeds it is a minimal effect on time while at high speeds it has a greater effect on time. As a body moves faster in space, it creates a sphere of increased energy density or region, and time is transmitted through the multi-node system much slower. However, this appears to be contradictory to our twin's paradox analysis, in which one twin sets out on a space ship at relativistic speeds only to come back to earth to find that the other twin has aged much more than they have. Because the earth bound twin continued to age at what we think of as a normal rate. In other words if their time actually slowed down for them their aging would not progress at a normal level but at a slower one. How is it that the space-venturing twin returns to earth and finds the other twin is older?

For the answer to this problem, we can again look to the science of aerodynamics for an analogy. When a plane flies at fast speeds, there is a boundary layer in contact with the surface of the plane. This boundary layer of air is actually moving with the plane, which makes its relative velocity to the plane zero. Matter moving through energy is a complex motion, in that the individual atoms are both vibrating and moving individually against energy particles, the object as a whole with all its atoms is also moving. So while our body of matter or spaceship in this case is moving faster, its apparent mass increases, because of the bow wave effect creating a cocoon of energy of a higher density then the ambient universal energy, similar to the higher energy density around a planet. The effect on time is that it slows down in this zone. However, a complex effect similar in principal the combinations of the Bernoulli's principle and the Venturi effect, seen in liquids or gaseous environments, take effect in the boundary region. The result is that the boundary layer becomes a low energy density discontinuity zone isolating the sequencing of time information inside the spaceship from the high-density energy cocoon around it and the ambient universal energy density. Mechanical movement by matter is less restricted in the ambient universal energy density then it is in the higher energy density around an object that is displacing energy. The higher the displacement factor of the object, the higher the energy density will be and the slower the mechanical movement in the region will be. This is why we see process like clocks and bio-activity seems to retarded and the inverse as we get into lower energy densities.

Our space traveler is held in their normal time before space travel, even as their spacecraft goes out through regions of reduced gravity, because they have become isolated in their energy cocoon that travels with them. What if they were to stay at a destination far removed from their point of origin, what would their time be? If we recall in our an ocean universe travel at C or

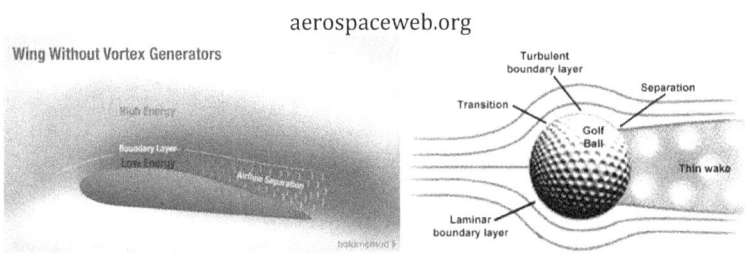

Fig. 13-1: Examples of low-pressure boundary layers and transition zones.
(BoldMethod.com, 2014)   (Scott, 2005)

greater speeds is entirely allowed, as the infinite mass issue has been resolved and the reason that we can only measure up to the speed of light is because we are using light as the measuring medium. If a traveler were to travel at the speed of light, he would keep up with the time horizon that was created when he first reached the speed of light. When he slowed to a lower velocity, he would again lag behind that horizon and find himself at the then current time horizon for his position in space. Our space traveler would find himself at a point in the future from where he was when he first reached C but not yet at the same time as the point at where he first reached C is now.

However, is he in the future or the past? Neither actually, our space traveling twin is still in the present. When he returns to earth and meets his twin who now seems decades older, he is still in the present, even though he appears to be a young man while his twin brother is now a well-aged elderly gentleman. One might think of the difference between a sequential read/write formatting system, such as early digital tape and disk drives versus random access formatting of our modern disk and flash storage systems. In a random access system, there is really no, before or after, there is just now, because the disk doesn't start at the beginning and read to the desired point it just moves to the information it needs now.

Of course, the whole twins time travel story/paradox only applies if the twin on the rocket is actually traveling faster through the ocean universe then the twin left on earth. What? When these mind experiments are created, they always are based on the assumption that the earth is the stationary frame of reference. Thus, it is why the twin in the rocket does not age while the twin on earth does? Should the motion be considered relative and they both age and both stay young relative to the other? The earth, like most of the things we see in the universe is moving through the energy of the universe like bobbers in the ocean. Besides just the increased density of energy caused by its displacement as matter, the earth also creates additional increases in the local energy density by the bow wave effect that it produces. So relative to an observer who is standing outside the universe or stationary relative to the whole universe, time on the earth appears to be traveling slower than for that stationary observer. For a stationary observer, one who is absolutely at rest relative to all things in the universe, the rate of time would go to infinity, as all things would appear to be happening at once.

The reason that the rocket twin in our mind experiment does not age faster than the earth bound twin is that relative to the universe the rocket twin is moving faster than the earth. Think of it in this way. If it is, pitch black out and we have two trains going in opposite directions and a car going along in the same direction of one of the trains. If an observer in the train going in the same direction as the car looks out and sees the car always in the same relative position to his train, then he will assume the other train is moving faster than he is. Even though, his train and the car may be moving twice as fast relative to the ground as the other train. If both trains are going in the same direction, but the other train and the car are going at the same speed, which is slower than the observers train, they will appear to stand still to the observer, who is moving away from them. If our rocket twin was able to plot a course in which he was moving a great speed relative to the earth but virtually standing still relative to the flow of energy around him then it is his clock that would appear to run faster and his twin on earth will have not aged as much when they are reunited. There is not a paradox with relativity; the apparent paradox exists only in the understanding of all the attributes of the problem presented.

Our present exists as a combination of the collective actions we are taking as body of matter, combined with the collective actions of the universe around us. Thus, our present includes the light reflected from our hand, the light hitting us from the moon, and the light hitting us from a star a million light years away. Contrary to popular speculation we are not looking into the past when we look at distance galaxies we are looking at light from a long ago event that is now present in our here and now. Suppose we were able to be in a massless ship, an information bubble if you will, and travel at the same speed as the time horizon from this instant, we will call, N=0. Would you always be in what the rest of us would consider the past (N=0)? It depends. If you think of a time event as a packet of information in a packet switched network then you would have to be either within in that packet to be a permanent part of the moment that packet records. If you were a conversely a continuous receiver of that packet of time information then you would continually view that event like an endless loop but you would not be part of that event it time. It would however be included in the recording of your sequential present viewing it. In the classical sense, you cannot go back in time and change the past. Events in the physical universe happen in a sequential mode, even if there are multiple strings of sequential events happening in parallel (not the same as the concept of parallel universes). However, it may be possible to change a moment in the past both in isolation such that the present goes unchanged and in a way that changes the present and subsequent moments leading to the present. We will discuss this in more detail after the following discussions on the intricate nature of the very essence of time.

Understanding energy's actual existence in the ocean universe opens the door for us to see how matter and energy interact and how that interaction is embedded in what we think of as time. If we transition from thinking of time as a physical part of our activity as beings of matter, then its existence as an information medium becomes more apparent. What if, we were to think of time more as a computer program, with energy as the memory or storage medium? Perhaps then, we could better understand how time is operating around us and within us. If we look to mythology we see several references to this concept; one most notably being that of the Alpha and Omega condition imposed by several god figures. Another is the concept of god being out side of time in that a day is like a thousand years and visa-versa. One last reference that we will explore is notion that if time were not shortened by a god, things would become unbearable for man. Can these concepts from ancient mythology be a part of a vague collective memory of humanity from the beginning of time?

We often think of Alpha and Omega representing two points of time, beginning and ending, yet what we do not always realize is that together they represent the whole. Metaphorically saying Alpha and Omega is the same as saying all time exists at once. If we think of computer programs as a sequencing of instructions, we have a very similar concept as what we have with the concept of time. Although a computer program exists in its totality, Alpha and Omega, it sequences through the lines of code that make up each of its modules or objects and that sequencing is always in one direction. From Alpha to Omega is also metaphoric for how the time program of the universe implements or runs. Now we cannot confuse loops in a program for the program running backwards. A loop is part of progress sequencing in one direction, which requires the rerunning of previous lines of code or objects. The routine that is called is not running for the first time all over again it is running a second, third, or n^{th} time again. The only way that the code could run the first time again, would be to reboot the computer clearing all memory and any stored variables; or what would be the thought of as the equivalent of the Big Bang.

Various gods talk about being outside of time and that time does not affect or bind them. A computer programmer is like a god to the program and all entities within the program in this sense. A programmer can enter and leave the program at any point and can change any variables of the program or even the very algorithmic nature of the program or natural laws if you will. To the programmer the program exists all at once, just as all the subroutines outside the program and called by the program exist all at once in memory and/or storage. A program can see any range of events by isolating and running only those modules of code desired. When we can understand that mythology, religious based or not, is a metaphorical way of describing real events, we can use the logic present in these myths, as a prototype if you will to model the real world around us. Perhaps in a different setting an ancient philosopher anticipated the functions a programmer (creator) would have to do to, much like a puppeteer in a play, and from this original philosophical explanation of the universe, someone created the many concepts of our gods.

Our last mythical reference has to do with the shortening of time by a god to save humanity from a very bitter existence. The first question we may ask is, "How would one shorten time and why would it become necessary to do so to protect mankind? This whole myth really begs an additional question of free will. In the context of the study of the nature of time, free will goes to the determination of, "Is time happening freely now or is it like a movie with every past and future frame firmly established and printed into the fabric of the universe?". Knowing if we can change the past or future allows us to understand better what the properties and rules of time are. Most of the theories of time that are taught and researched today have their foundations either in the distant past, or in the more recent past of the late 19th and early 20th Centuries. Therefore, our contemporary understanding has its foundations in a very antiquated technology base.

For example, one of the most popular analogies of time is that of a movie film with numerous sequential pictures actually permanently arranged on a piece of transparent film. Immediately we are intellectually boxed into a antiquate vision of how time might occur. Today few of our brightest students of physics would envision a literal movie film to describe either a movie or time. In the digital age, our understanding of information sciences is beyond even the imagination of the giants of intellectual thought during the waning days of the 19th Century and the beginning of the 20th Century. Today it not only more accurate but much easier to describe time in the context of digital media, where the content of the image is stored as data in a virtual state rather than as hard images on physical analog medium of see-through film.

When movies were stored on rolls of cellulose acetate, film movies had one ending, the one the director decided on. Today movies stored as virtual digital data on Blu-Ray disks, offer a variety of endings to satisfy the cinematic taste of a broad spectrum of viewers. Some of the latest technology allows the viewer to be the director and actually describe the type of ending that will be shown, while this ending is virtually created from material on the Blu-Ray. We have not even addressed the world of modern computer gaming technologies with Computer Generated Imagery (CGI) and artificial intelligence based software (Graepel, et al., 2008). Today those exploring the concept of time, who have not had their vision boxed in by outdated theory based on antiquated technology analogs, can see time in the context of virtual reality where not only are there many time lines possible, but each is easily and readily changeable.

In essence, we need to think of time in the context of contemporary information science, where the physical universe is the computer hardware and time is the software that that controls, sequences, and records all the events that take place in the universe. With this understanding of the function of time as universe's "software" the ability for events to play out in an ordered fashion while the concept of free will and/or chance and randomize coexist, just as in any advanced computer game or viewer interactive movie. In the ocean universe, past, present, and future all exist at the same time and the energy particles are the storage/network devices and processors, in a similar sense to a neuro network, such as the neuron network of our brains.

Today we are still only beginning to understand the complexities of natural information systems such as the human brain, with millions of connections and nodes. In information science, we are now working on second order neural network architectures. These systems while advanced beyond the imagination of practitioners a generation ago are still only beginning to tap the potential abilities of information manipulation and control. Today we are dealing with systems such as Synaptic, which has capabilities such as Multilayer Perceptrons, which is a feedforward neural network (FFNN) model. Multilayer long-short term memory networks (LSTM), a type of Recurrent Neural Network (RNN), which unlike a FFNN, allows for arbitrary inputs. RNN make it possible for systems like those that can learn to read handwriting. Hopfield Networks, which mimic not only human memory, but also serve as a content addressable memory, may prove to mimic some of the functions of time (Cazala, 2015). The technology has come a long way since this author was managing the development of advanced Research and Development projects developing Generators of Code Generators. These generators were being designed to convert native language into a software automation tool that could then develop other software tools that could write code in specific languages for domain specific programs. The idea was to give bankers, lawyers, doctors, plumbers or any other domain expert the ability to describe their work in plain language, and then be able to create useful programs for their business without a human writing the code. This effort was done under the Software Design for Reliability and Reuse (SDDR) project sponsored and funded by the United States Air Force Material Command (Bell, et al., 1999).

One of the most significant questions relating to time in physics is about the ability to travel backwards and forwards in time. Recent work by Yakir Aharonov, a physicist at Tel- Aviv University, and his colleagues suggest that what is happening in the present today can actually determine what happened previously in the past. Based on the concepts of "Two-state Vector Formalism", work has been done at the particle level to show that what ultimately happens to a particle at a present point in time can actually change its state at a past point in time (Aharonov, et al., 2012). This concept is completely consistent with the ocean universe and the notion of time as a part of a natural digital SCADA like system.

The first way you could influence the past, is if you were to replay a period of time you could, in that replay (play in the sense of a video game rather than a movie) you could make different choices and/or effect different outcomes of events. However, the replay is a second running of the scenario and not a change in the original time record. Like a second iteration of a computer subroutine. The problem will be that all other calls for that subroutine output must be reinitiated if the changes, from that modified subroutine (moment in time) are to update across all timelines. For example, suppose someone died in the past and you were able to go back in time to save his or her life. If their previous death was to have happened in a subroutine then that data would show them still alive but they may appear to simply vanish in other time segments because their continued existence was not updated across other routines. This approach is most consistent with the concept of time travel as portrayed in science fiction movies. The problem with this concept is how we don't see time changing more often? For example if the future already exists, which it must for us to be in the past today, why are there so many social abnormalities still in our history. For example, would not a person want to come back and prevent wars? If so, why is it that Pearl Harbor still exists in our history books? Why is the Titanic still at the bottom of the North Atlantic? Why is Jesus still recorded in history as being crucified?

A second possibility would be to browse/search the event database and open a record for a given period of time and just make a change to the data set. This would alter the data stored for that time-period and later calls on that data set would invoke the changed data rather than the original outcome. One problem with this would be, for example, if you changed the record of what happen at Pearl Harbor, is how you would explain the missing ships and people in subsequent time data sets. You would still have to update all files and variables of that data to reflect that Pearl Harbor did not happen and account for the people and the ships.

The third option for time travel is to revisit a moment and time to watch it be redisplayed, like watching a rerun on TV. This would be most consistent with the concept of a rocket traveling fast enough as to go backward. If a time event horizon is disseminating outward at the speed of light and an observer could get ahead of that horizons propagation through the multitude of nodes, then the observer could re-witness that event. However, it would be like watching a movie, because the original event happened somewhere else. Think of a shortcut icon on your computer desktop. When you click it the program or file it is linked to activates, you can see that activation, but by changing or deleting the shortcut icon, you cannot affect the actual file or program. This is probably the safest form of time travel for us to participate in.

How do we now address the Grandfather paradox? This paradox involves a person going back in time and somehow preventing his grandfather and grandmother from producing their father or mother. What would happen to this person once they prevented their parent from being born? Classically the most popular scenario of the grandfather paradox is that the person would not be born, would not therefore be able to travel back in time to prevent their parent's birth, therefore the individual would be born, and the loop would continue until a loop occurred where they decided not to travel back in time. Of course, a number of other theories exist such as the Novikov self-consistency principle/Temporal Modification Negation Theory, which states that nothing can happen in the loop that would prevent the time travel from happening in the first place. Or, a theory that is fast becoming popular, because it allows for any desired definition to fit, is the Parallel Universe theory, which requires each case of time travel to result in a destination that exists in a parallel universe. As strange as this theory may sound, it does have an abstract sense of reality to it.

All computer programs have a sequential logic of some type to them. We may mask this sequencing in the form of parallel processing, object calling, subroutine runs, and other forms of design that may split-up the workload into separate pieces but, within each of these separate module, threads, or objects the software continues to run in sequence internally. If we think of time as the digital infrastructure of a software intensive system, our universe, then we must understand that all events happen sequentially. As we are part of the overall system, we are bound by the same sequential rules. Once we have gone by a point in the sequencing, then just like in any other software system, we cannot go back to it without rebooting the system and running from scratch again. You may think we can rerun routines and to this the theory and facts holds true. The operative word here is "rerun" or in other terms, "loop" or "goto" or "for next" or "function call" or "jump". Regardless of the terminology, the software system is going to track a "next" function to determine what iteration it is in, thus maintaining sequentially. What this means is that although a program is repeating a previous set of instruction sets it is doing so in the sequential present not the past. The software keeps track of iterations as part of its operating instructions and when the iterations are complete then the program continues to go forward in the next set of instructions. It is interesting to note that even in first generation software the data sets would still keep track of iterations, even though the original data record may be over written to conserve storage space. While in more advanced systems the importance of meta-data is incorporated so that data on iterations of processes is maintained and data sets are not overwritten but rewritten in new areas of data storage.

With our current understanding of information sciences, which is far from compete, rather than early 20th Century movie film technology, we can see that the grandfather paradox is virtually a non-issue. In a sequential system if one was to revisit a past event they would be doing so in the present tense. In other words, doing a call or jump to a previously executed module in software takes place in the sequential now, not in the past. To visit your grandfather in the past you would call your grandfather's subroutine/object and enter it in the now, not the past. If you change things in that routine they would not affect the original running of the routine only the current running of the routine in the present. If you killed your grandfather, you would be doing it in your present after you were born not your past (think sequential code execution in a meta-data environment). Of course, you would progress forward in the sequential execution of time, when you exit the "grandfather subroutine/object" in the current time, but you would have a past of grandparents that did not really exist. Imagine if you worked with the CIA and could erase all records of your existence, birth certificates, Social Security Numbers, photos, driver's license, friends and family memories, etc. This would be you, alive but without a past as you know it. Again, if you were a computer gamer, used to modern interactive games with characters whose actions are controlled by artificial intelligence, you would be familiar with this type of interaction. You would also be able to, in some games, interactively adapt the game to new scenarios on the fly based on player preferences.

In order to move forward to the next levels of scientific understanding, which will be able to propel us to new heights of social-technological achievement, we have to let go of scientific theories base on antiquated technologies and analogies using metaphoric imagery that simply is not true or possible. We need to revisit out understanding of time and space-time from a perspective of modern advanced information theory and science, recognizing that time is the software of the largest software intensive system ever created, our ocean universe.

Chapter 14 – Experimental Flaws

The Parallel Mirror Experiment

Fig. 14-1: The flawed parallel mirror explanation of time dilation using a
light beam bouncing between two parallel mirrors (Science TV, 2010).

Theories are valuable tools to explore the universe around us
but unless they can be tested, evaluated, and validated they are of
little value, except as a mind exercise. Even with all the latitude
that is afforded, theories in the pursuit of new understandings in
theoretical physics, a certain sense of consistency and reality must
be maintained in the fabrication of experiments if they are to be
valid. With this in mind, before we look at actual experiments that
could be performed to either validate or invalidate any of the
postulates of the ocean universe theory; we will demonstrate
some fatal flaws in some of the more popular mind
exercise/experiments being proliferated in modern physics.

One of the most prolific of erroneous experiments involves two parallel mirrors with a light beam source that shines through one mirror and then bounces off an opposing fixed mirror parallel to it. When the apparatus for this experiment is configured, there are two parallel mirrors spaced a known distance apart. A light is flashed from one mirror to the other and the time of the round trip is measured. As long as the experiment is relatively stationary, the light beam will bounce back and forth between the two mirrors at a constant rate, which can be determined by the distance between the two mirrors divided by the speed of light. In figure 14-2, we can see this layout. If the distance between the mirrors is 10 meters and with light traveling at ~299,792,458 meters per second it will take about 6.67128E-08 seconds to make a round trip from the lower mirror to the top mirror and back. For the purposes of this experiment, we will say the light beam (a single flash) is 3.33564E-10 seconds or approximately 10 centimeters long.

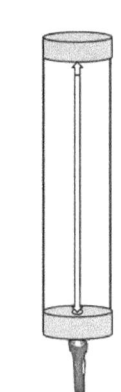

Fig. 14-2: Basic Setup of Parallel Mirror Experiment

The point of this experiment is to try to demonstrate how and why time would appear to slow down for an observer standing still while the time for an observer on the moving platform would stay the same. It is loosely based on Einstein's throwing a ball from a moving train postulate. In this postulate, the observer on the train sees the ball fall straight down while an observer on the ground sees the ball fall in curve in the direction of travel of the train. Basically, this parallel mirror experiment is supposed to demonstrate the same basic principal only substituting light for the ball.

The experiment, if it could actually be conducted in the physical reality, would be executed by moving the mirror mechanism at a constant relativistic velocity along a vector that would bring it left to right in front of our stationary observer. At normal speeds, that is speeds that we are used to experiencing if a pulse of light 3.33564E-11 or 10 centimeters long is flashed from the bottom mirror (point a) to strike the top mirror 10 meters away (point b) and return in about 6.67128E-08 seconds to the bottom mirror (point c). At speeds normal to us, we would have been able to move the apparatus and infinitesimally short distance. The measure of arc between the start point (a) on the lower mirror and the return point on the lower mirror (c) would be infinitesimally small. While the time from point (a) to point (c) might be able to be measured it will curiously show that the time divided by the distance traveled will resulting in the light traveling at C. This is because as we discussed earlier we can only measure how far light has traveled and how long it took to travel when it actually impacts into detector devices such as an eye or photoelectric cell.

Here is what you will not see in this experiment. You will not be able to see from the side, as shown in the previous figure, the actual light beam moving between points. When we see a laser beam, what we are seeing is light that is bouncing off smoke or dust and redirected towards our eyes. Today with modern Femto-Photography with cameras operating at relativistic speeds, it is only possible to photograph light beams passing though smoke or dust. No observer could ever possibly see the actual light from the side of the device, no matter at what angle the observer stood. A light beam traveling in a vacuum, as defined in this experiment, is not visible from the side. We only see the reflected light not the actual laser beam itself. Light travels in straight lines, even when those lines are bent around an object by gravity, the light is still considered to be traveling in straight lines. Thus, our laser light would emit a narrow beam of light that would travel directly from point (a) to point (b) of the experimental device with none of it radiating outward from the zero-angle-path of propagation. All light is like this even our flashlights; only in our flashlights, the design of the bulb and the reflector creates an infinite number of straight lines of radiation outward in an infinite number of angles from the zero-angle-path.

While many of us have seen light swords on Star Wars, and laser beams in laboratories or rock concerts, we might say this is not apparently true because you could see the beam of light from the side. Actually what you are seeing is reflected light from small particles in the air when the beam traveling in a straight path perpendicular to your line of sight hits these articles and is reflected along a new vector directly at your eyes. As previously discussed and explained we can only see light that is directed at our eyes and we can only see or detect it when it actually reaches our eyes/senor not any time before.

Now what if we were to rerun our experiment only this time we were to move the test apparatus left to right in front of our stationary observer (Fig. 14-3) at a speed of 5% C or about 14,989,622.90 meters per second. What will be our results be if we again flash a beam of light of a length 3.33564E-11or about 10 centimeters long from point (a) on the lower mirror to point (b) on the upper mirror? Once again, we would not see the beam of light if our experiment was to take place in a vacuum, and if it was

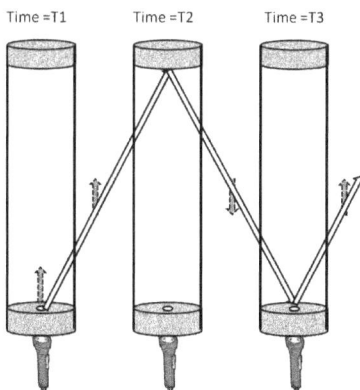

Fig. 14-3: Parallel mirror experiment as traditionally depicted. Long hollow arrows represent supposed path of light as seen by ground observer. Short dashed arrows represent relative motion of light to mirrors (Straitt, 1980-2015)

not in a vacuum the light we would be seeing, would be reflected light off the particles of matter, not the original light beam itself. If for the sake of the critique of this experiment, we imagine we could somehow see this light from the side and if we could actually comprehend something happening this fast, what we would see would further confirm the complete fallacy of this experiment's proposed proof of time dilation.

As the following diagram shows, if we are to move the test apparatus from left to right at 5% of the speed of light (C), 14,989,622.90 meters per second, we will actually move the parallel mirror apparatus from the path of the light beam long before it can ever reach the top mirror to be reflected down. The faster we go the further away we get from the path of the light beam. If we were to have a left to right velocity of C or 299,792,458 meters per second we would need a set of mirrors a little over 10 meters in diameter for the light beam from the lower mirror to reach the upper mirror rather than going off into the universe. However, you could never bounce the light off of the

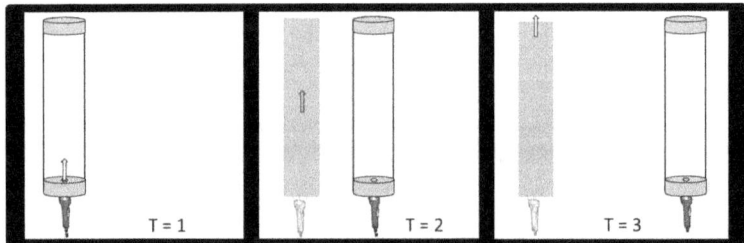

Fig. 14-4: Parallel Mirror experiment as it would actually occur if the apparatus is moving left to right at a velocity of 5% of C.

center of the upper mirror if you fire it directly up from the center of the lower mirror. Light does not behave like a ball falling from a train window as Einstein discussed, it is because the light does originate from the moving platform but from a fixed point in space that the moving platform with the light generating device is passing by. Light waves propagate relative to a moving platform like water waves propagate in a pond, if you hover over a pond and drop stones into the water. Regardless of the speed of the moving platform, the waves will propagate out from where the stone hits the water, and each stone will create a new wave front independent from the previous. The duration and amplitude of each wave front will be dependent on how much force (mass and velocity) that the stone hits the water with, not on the speed or direction of the moving platform. This is how light is created, in a digital pulse format rather than a continuous analog format.

Now we have made some pretty dramatic arguments, tearing apart a long standing model of time dilation with some diagrams and complex explanations, as well as some simple mechanical models. We can back this up with some real world evidence. Photographs of light behaving in just the way we have described it within this book.

Some 30 or more years ago when I first started describing this model of the ocean universe and the mechanics of light technology did not exist to verify the theories set forth herein. Today, that technology exists in the form of Femto-Photography, which is able to capture events at one-trillion frames per second.   Femto-

Photography, developed at the Massachusetts Institute of

Technology by Professor Ramesh Raskar and his team, is able to capture finite light beams emitted from a green laser. Of course, as we discussed earlier we cannot see light directly from the side but

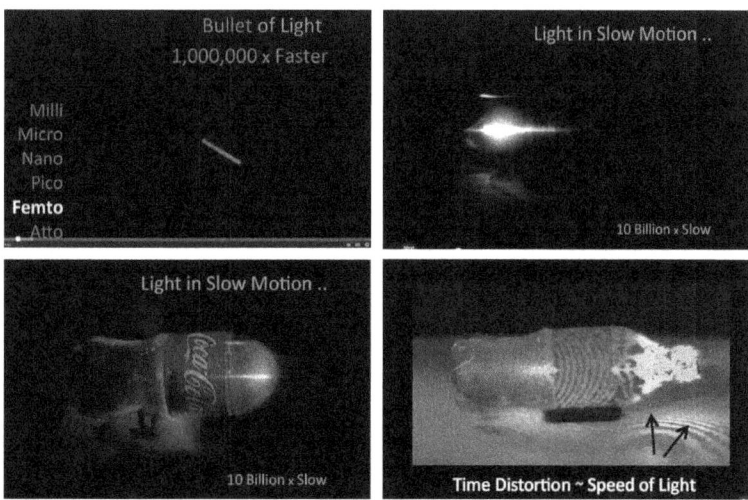

Fig. 14-5: Actual stop frame motion of a beam of light (upper left), a beam of light shot through a coke bottle filled with water (upper right, lower left), color enhance picture of wave fronts from reflected light from a beam shot into Coke bottle (lower right) (Raskar, 2012).

we can see reflections of it as it pass through fine particles of dust or smoke. What we see from this demonstration is that light beams are finite. That is they have an absolute beginning and an absolute end. They travel in a straight line and they interact with matter as they come in contact with it. In the figure 14-5, you can clearly see a finite beam of green laser light that has a length of 1E-12 seconds or .3 millimeters in length. In the upper left picture is the beam of light entering into the base of a plastic Coke bottle filled with water. The diffusion of the light beam in the water can be seen as the beam penetrates the further towards the neck of the bottle. In the lower left is the light beam as it is about to hit the cap of the bottle and explode into individual photons (Raskar, 2012). We will discuss the last photo after we look at these slow motion pictures of water drops hitting a surface of milk in figure 14-6.

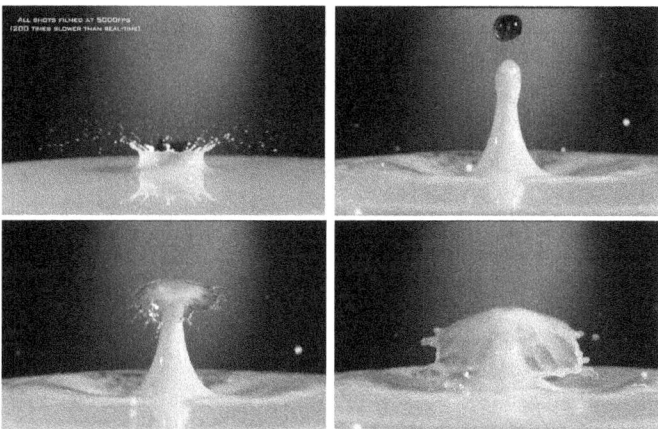

Fig. 14-6: Two drops of water hitting a surface of milk. Note how the first drop does not just merge into the liquid it creates a scattering patter and loses some of its mass. In the top right and bottom left pictures the second drop is about to hit the rebounding first drop. In the lower right, the second drop is splattering out in inverse cup like sheet of liquid (Free & Gruchy, 2011)

In our water droplet figure 14-6, we are seeing two water drops hit a surface of milk, filmed at 5,000 frames per second. In the upper left picture we see that the first drop has just hit the surface and has pushed out a cup like splash of milk. Notice that there is a distribution of droplets spreading upward and outward. What cannot be seen in the still picture is that the droplets are actually moving outward faster than the initial drops velocity when it hit the milk surface. In the upper right picture, we see that the initial droplet is rebounding from the surface and is bringing some of the milk with it. This phenomenon is called the Coalescence Cascade, and had their not been a second drop colliding with it, would have continued until all the water in the droplet was dispersed into the milk surface. In the same picture, we see the second drop about to impact with the coalescence from the first drop. In the lower left picture the second drop and the coalescence from the first drop have impacted each other and that impact has created a new coalescence. In the lower right picture, you can see that coalescence from the impact of the second drop is spreading out but the cup shape is inverted down wards rather than upwards like the original drop.

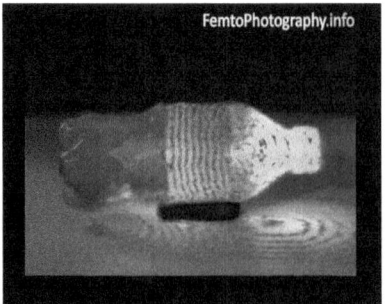

Fig. 14-7: Reflection on table artificially inverted with Software (Raskar, 2012)

Back to our Femto-Photograph, Fig. 14-5, of the light

beam impacting a Coke bottle and looking at the picture in the lower right we see the wave pattern of the light coalescence. What is noted by arrows is that the shape of the waves on the tabletop was opposite of what the researchers expected it to be. The researchers attribute this reversed direction of the waves on the table to an error in the software integrating the image segments due time dilation. To correct for this perceived error they created an algorithm to modify and invert the reflected waves on the tabletop, as shown in figure 14-7.

This demonstration shows that light beams can interact with themselves, as is actually evidenced in the lower right picture of the modified image showing the wave fronts being concaved away from the bottle. Experiments performed by this author with a laser pointer and a Coke bottle show that the reflected light does tend to be concaved away from the bottle as in the unmodified image. Although this effect is associated with time dilation of the fast moving camera, an alternative reason for their appearance is light tacking. As we explained earlier, tacking is a sailing technique, which allows a boat to sale faster than the wind and it is based on vector addition. It requires that boat to sale at an angle to the true wind and have a resistance with the water against its hull. The same thing can happen with light waves being pushed along by other light waves at obtuse angles and encountering a similar drag force as a hull of a boat would.

As the first part of the wave reflects outward from the area of the cap, it creates a pressure ridge, when the next part of the wave hits the cap it is also reflected out, but it presses against the pressure ridge created from the previous part of the reflected wave. While the force of the rest of the incoming wave continues to push these earlier reflected-wave-segments both outward and against the pressure ridge it creates an effect much like that occurs when a sailboat is tacking against the wind. The effect is described by the Beta Theorem, which is also known as the Course Theorem (Gilbert, 2012), results in an apparent force (wind if it were a sailboat) acting on the reflected wave with an additive effect of two vector velocities. Just as a sailboat tacking downwind achieves an increased velocity so would the reflected light waves. Each light wave (which is propagation effect within the energy medium) would be slightly accelerated while under this influence of the Beta Theorem effect. However, once the second and all following reflected waves accelerate past the end of the pressure ridge they lose the additive velocity effect and drop back to normal speed. The cascading effect continues until the end of the oncoming wave is reflected by the cap and because now there is no more wave segments coming behind it to create the Beta Theorem effect it is not accelerated as the previous parts of the reflected wave.

Thus, the beginning and end of the wave arrive at the tabletop last while the middle part of the middle part arrives first in reverse order. The front of the wave when reflected is not subjected to the Beta Theorem effect because it there is no pressure ridge from a previous reflected wave to for the last wave to press against. If the force of the original incoming and the resistance of the pressure ridge caused by the reflected wave remained linear, the effect on the reflection on the table would be linear so that the waves would be straight and parallel. The forces and resistance are not linear resulting in the curve of the wave patterns being bowed into the Coke bottle as seen in the unmodified photo. Ultimately, there is no time dilation, rather a normal wave propagation event, influenced by an effect similar to the Beta Theorem effect commonly used by sailing vessels.

Young's Double Slit Experiment

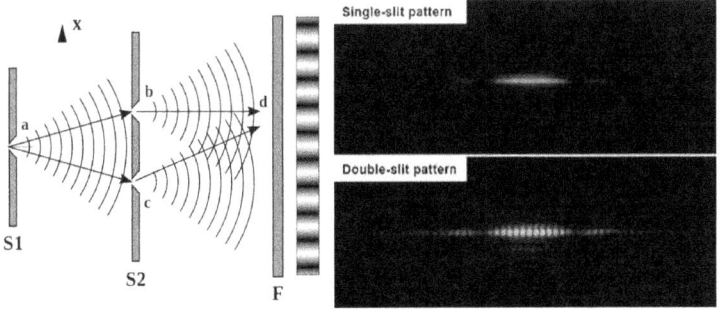

(Stannered, 2012)   (Williams, 2011)
Fig. 14-8: Young's Double Slit Experiment Configuration.

In 1799 Thomas Young presented a paper to the Royal Society proposing that light was a wave propagation rather than as the

Corpuscular theory of light proposed that light was made up of little particles that traveled in straight lines (Young., 1807). While Young was convinced that his experiment definitely showed that light was a wave action, later Quantum Mechanics theorists would use the same experiment to show that light was actually a particle action and that the Double Slit Experiment verified this. In both the pre-particle and now particle theories it was shown that unique pattern found on a screen behind the double slits always occurred, except if you tried to take a measurement. It has been found that if any device is set up between the plate with a single slit and the plate with the double slits will cause the distinctive pattern found on the imaging screen to change to a pattern found behind a single slit plate.

With the contemporary emphasis on the particle theory of light in a Quantum environment, various active and passive devises have been used to try to determine a particles path after exiting the single slit and making its way through one of the slits in the double slit plate. In each case, the familiar interference pattern that is normal behind the double slit plate disappears when even the most passive of monitoring devices are used (figure 14-8). Theorists generally associate this phenomenon with what is called the Measurement Problem in Quantum Mechanics.

The flaw in the experiment is that the theorists do not take into account the nature of energy in the ocean universe. When we think of a wave in water, it is easy to both visualize the wave action in a pool, as well as, the molecular interaction between the water particles. One could easily see that by spraying a garden house into a ripple in the pond hoping to catch a reflected wave front from the spray to monitor the ripple would be an exercise in futility. Using any form of electromagnetic radiation to sense a wave traveling between the plate with the single slot and the plate with the double slot, is like spraying a fire hose into a swimming pool and expecting not to disrupt the gentle ripples from tossing a pebble into the water. Likewise, if it is assumed it is particles or if it is actual particles like electrons passing between the plates with the slits and you try to measure the particles with any electromagnetic wave then it would be like spraying a fire hose into a formation of bobbers floating in a pool.

The ultimate flaw in the logic of the experiment is that the particles can somehow deduce that you are looking at them and change their behavior. When we interject a wave or particle stream into the space between the two plates we are disturbing the original wave/particle pattern propagating out of the single slit. In essence, the two slit plate now becomes a two single slit plates side by side to each other. This does not mean we could not develop a measurement tool that would allow us to look at the wave activity and maintain the diffusion pattern on the screen. The probability equation is probably not a part of the process. In order to measure the wave effect without converting it into an apparent particle beam we need to use a measuring medium other than light (electromagnetic waves) or matter (particle beams) to do the measuring. What we are left with then are two possibilities time and gravity. In the context of this particular book, the discussion of how to use time to physically-measure a light wave scattering pattern may be premature. However using the term "gravity waves" is an appropriate discussion for this publication.

As discussed earlier, gravity is a force caused by displacement of energy particles. We also discussed that although matter is the principal candidate for such displacement, it is possible to have a situation where energy can cause significant displacement of energy resulting in an apparent mass. When this occurs there will be gravity waves/forces produced by the energy just as with matter. An analogy that may be helpful is to think of the difference between the water at the ocean surface and the water at the ocean depths. The pressure at the bottom is such that the water if it could be maintained at that pressure and weighed would weigh more, or have an higher apparent mass per cubic unit of volume than the same volume of water from the surface. What if a wave was passing over an area of the ocean bottom and that wave was of a significant enough height to cause a measurable difference in the apparent mass of the water at the bottom of the ocean. If you could somehow interact with this change in the apparent mass of the deep ocean water, you would be able to determine quite a bit of information about the wave passing overhead.

It is proposed that you can do the same thing with energy in the ocean universe by measuring the density change or gravitational wave caused by the light wave passing though the space. Since this would be a truly passive form of observation, it would not disrupt the wave passing between the two plates in any significant way. How would one measure this change in density? A proposed method is outlined in the appendix on suggested experiments. It involves using a body with an apparent mass generated artificially and then monitoring the changes that occur to the mass as the waves pass by. Much the same as you would use a pressure gage to monitor the changes in the pressure of water at depth when a wave passes overhead.

Fig. 14-9: Floating Object in glass of water Expirement (Shanker, 2012)

Surface Tension Experiment

Although this flawed physics experiment may not seem directly related to the physics problems of the ocean universe, since we discuss ping-pong balls floating on water earlier in this book we need to address the flaw in this popular experiment. The experiment is arranged by placing a ping-pong ball in the center of a glass mostly filled with water. The ball will not stay in the center and will always move to the edge. When the glass is filled completely with water to the point where the water is about to overflow the edge of the glass, the ball will not stay on the edge of the glass and will always go to the center of the glass.

In figure 14-9, a diagram of a floating wine cork (ping-pong balls work also) experiment, we see the representations of the force that is commonly associated with moving the ball towards the edge of the glass. The concept is that the forces of surface tension acting unevenly on the floating object cause the object to be forced to the edge of the glass when the glass is not completely

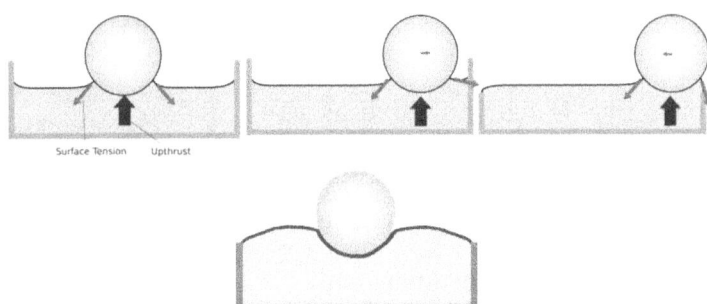

Surface Tension   Upthrust

Fig. 14-9: Popular experiment attempting to show surface tension effect of water on an object floating on water in a partially filled and fully filled glass (University of Cambridge, 2010).

full and to be forced to the center of the glass when it completely full. However, the true answer to this phenomenon is that the floating object is just seeking the highest point on the surface of the water. It is a gravity related issue, not a surface tension one. When the water surface tension and capillary action is causing a rise around the outside edge of the glass, then the highest point on the surface of the water is at the edge and gravity is causing the lightest object to float to the highest point. When the water is filled right to the top rim of the glass, again surface tension and capillary actions cause the water to convex upward and the highest point is at the center of the glass. Basically, it is like watching a float rise up and down as waves pass underneath it (Shanker, 2012).

While the ball or other floating device may be floating due to displacement and surface tension its movement to the side of the glass or the center of the glass is a direct result of gravitational stacking or separation of materials by their mass.

Speed of Light

Although we discussed the measuring of the speed of light in earlier sections, it is important to address the mechanics of the experiment here once again. From the Michelson–Morley experiment of the 1880s through modern times with laser light sources and advanced sensor systems. The experiments for measuring the speed of light relative to a moving observer have all suffered from one fatal flaw. That critical flaw is the use of light as the yardstick to measure light. This flaw is as simple as it sounds, and has eluded some of the greatest minds in science for centuries. Any measurement of light, which involves the detection of light by the observer, visually or through instrumentation, requires the

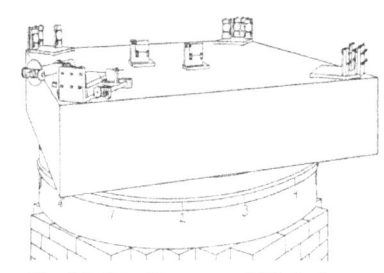

Fig. 14-10: Diagram of Michelson–Morley experiment as reported in 1887 (Michelson, 1887)

light to have physically reached the detector. Since it has traveled the distance from its source to the detector then the speed is always going to be measured as the time it took for the light to travel the distance from source to detector, regardless of the motion of the detector. Even if the source is moving, the speed of light is still measured from the point in space that the light excitation source was at when the light originated not from the excitement source itself. In other words unlike a bullet that is fired from the relative position in the moving barrel of the gun, thus the velocities are multiplied. A light beam originates at a point in the ocean universe as a wave in the ambient energy of the universe, not from a relative

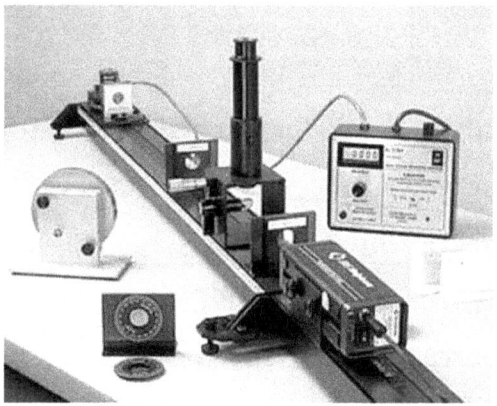

Fig. 14-10: Typical modern instrumentation for conducting experiments to measure the speed of light in a classroom (PASCO, 1996-2015).

stationary point on the flashlight/laser. In the next chapter on proposed experiments, we will explore in detail how to measure the speed of light accurately, relative to a moving observer and a stationary observer. The method proposed will eliminate the need for trying to adjust for time dilations and will provide accurate measurement regardless of the motion or non-motion of the light source and/or the observer.

Therefore, weather you are using the original Michelson–Morley experiment, the Kennedy–Thorndike experiment of the 1930s, or the more modern approach of using optical resonator/optical-cavity configuration, your design of experiment suffers from the same fundamental flaw. If you measure any wave speed, light, air, or water using only the time of departure from the source point in the medium and the time of impact at the sensor, the speed will always be constant, regardless of the motion of the sensor or the excitation devise. All waves are a property of the medium they propagate in and the speed of a wave in a medium is fixed by that medium, independent of the motion of the observer.

The reason we do not run into this flaw in measuring other wave phenomenon, besides light, is that we can use light as a ruler to measure the ongoing progress of the wave front. Light is traveling so much faster than any other wave action we are measuring that a light beam used to observe other phenomena, is able to reach out to an oncoming wave front or other object and then bounce back to the observer before the measured wave has propagated significantly.

Chapter 15 – Proposed Experimentation

Experiment 1 - Validating the additive speed of light

Before addressing any other properties of light and the ocean universe, we should first propose an experiment that will validate that light is in fact, a wave propagation in the energy medium of the ocean universe. This experiment will show that although the speed of light is generally constant, it is not limiting speed to travel through the universe.

Unlike prior experiments suggested and or performed, the nature of this experiment will validate that an observer's velocity is additive to the velocity of light. As we previously discussed and seen through femto-photography, light travels in beams of finite length and breath. In normal propagation, these beams move out from the point of excitement in all directions forming a wave. However when focused with a laser these beams can move as discrete packets of propagation in one direction. The length of the beam will be equal to the time of the excitement action/flash. As this beam is traveling at C it will take as long to pass a given point as was the length of the flash/length of the beam. Thus, a three nanosecond beam of light will take three nanosecond to pass a point in space or to be completely absorbed by some type of receiving devise. If we were to embed a data string within the beam such that the data began on nanosecond after the beginning

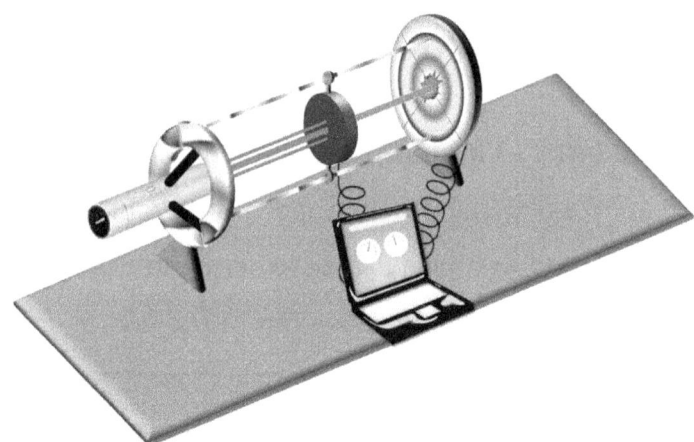

Fig. 15-1: Setup for an experiment to measure relative additive velocity for a moving observer/sensor and a beam of light (Straitt, 1980-2015).

of the flash and the last data-point was imbedded one nanosecond before the end of the flash. Our whole data package should be received at the sensor within one nanosecond of its arrival at the receiving device, unless there is a change in the relative velocity between the light beam and the receiving device.

To accomplish this, verification it is proposed to fire a laser beam of a very short length, which is to be determined the physical dimensions of the test bed, at a target that is designed to allow some of the beam to pass through the target uninterrupted while capturing some of the beam to receive the data set. Both the stationary solid target and the moving donut like target are attached to a computer based timing and control system with flexible wires of the same length. We use a wired rather than radio based monitoring communication so that no matter where the target is in its path the signal from the moving donut target will travel the same distance along the wire to the computer as the signals from the stationary target. The donut shaped movable target is mounted on a set of rails in such a way that the target can be accelerated to a high rate of speed towards the laser at the end opposite the stationary target.

With the donut shaped target stationary, a 3-nanosecond pulse with a signal is fired from the laser towards the movable and stationary targets. The timing of the speed of light can be calculated by the time it takes to receive the whole data signal. This timing will include the start time of registering the first waves of light and then measuring the time interval between the arrival of the first wave/photon of light and the first digital data point embedded within the beam. The same is done for measuring the time between the last digital data point embedded in the light beam and the last detectable wave/photon of light. In addition, the time it takes to receive all of the digital data points within the beam will be measured. With both the targets being stationary, the time recorded for both targets should be the same (within the acceptable limits of design of experiment variations). Further, the timing of receiving at the target should match the timing of transmission at the laser of the first wave/photon of light and the first digital data point embedded within the beam. The same is done for measuring the time between the last digital data point embedded in the light beam and the last detectable wave/photon of light. In addition, the time it takes to receive all of the digital data points within the beam will be measured. With both the targets being stationary, the time recorded for both targets should be the same (within the acceptable limits of design of experiment variations). Further, the timing of receiving at the target should match the timing of transmission at the laser.

Now the experiment would be performed again only this time the moving target will be accelerated to a predetermined relativistically significant speed. When the movable target reaches this speed the same length beam of light with the same digital data beam will be fired from the laser. Upon performing, the same timing measurements the expected results should show that: a) The stationary target records the receiving of the light wave and signal with the same timing as before, which will be consistent with the timing of the transmitted light and digital signal. b) The moving target records that a shorter time of reception of the light and digital signal then that recorded by the stationary target versus what was actually transmitted by the laser. The shorter time of reception of the light and data by the moving target will correspond directly to the combined velocities of the light beam and the moving target.

One of the criteria that has been long established with the physics community is that any experiment that is conducted to determine the speed of an object relative to the speed of light must not fall into what is commonly known as the "Lighthouse

Paradox". This paradox arises because of the effect of rotating a powerful searchlight beam or laser-beam transverse across a very large arc, relative to the light source, appears to moving along faster than the speed of light. To address this paradox, the physics community has first, generally agreed that the light spot/laser dot is not a "thing" in the context of say a piece of matter. Secondly, it has been established that the in order to be a valid confirmation the object/thing moving would have to be able to convey some sort if information from point A to point B along the path of movement. A spotlight shined on point A at some great distance and then rotated at great speed to point B at some great distance, will not be able to convey any information about point A to point B. Of course, in both cases it will convey information about its origin to both point A and then to point B when it shines on them, but this will of course occur at the speed of light.

In the experiment proposed herein, the validity criteria to be able to transmit information faster than the speed of light is met in two ways. First, the length of time that laser light is shining on the moving target is less than the length of time that it is shining on the stationary target. So that if you were to send multiple flashes in the form of a Morse-Code message of dots and dashes, the moving target would receive the message faster than would the stationary target. Albeit that the moving target dots and dashes would be proportionally shorter in duration, and directly related to the relative speed of the target to the light beam, on the moving target than they are on the stationary target. Second, the digital data set is received at a faster speed on the moving target when it is moving then when it is stationary and faster than received on the stationary target. The increase in the speed of reception on the moving target is consistent with and directly proportional to the relative speed of the moving target to the beam of light.

Variation on Experiment 1 – Case A

A variation of experiment 1 would be to replace the stationary target with another laser beam pointed along the path of the moving target and directly at the first laser beam. Additionally the moving target would have sensors affixed to both sides and a femto-photography camera would be mounted perpendicular to the path of the laser beams, to capture both laser beams (traveling through smoke/fog) and the moving target in a single frame. Of course, as a femto-camera provides a computer generated composite image from multiple data input sets, the software of the camera would have to be configured to just display the image as it is actually being sampled, and not artificially corrected for perceived time dilations, as is commonly done to adjust some images to conform with popular conceptions about time dilations.

This time when the experiment is run the expected results would show that the laser beam flashed along the path of the moving target will be converging on the target at a speed greater than the speed of light. The laser beam flashed in the same direction as the moving target will be converging with the target at a speed slower than the speed of light. The femto-photography will capture the distance and times in a manner that they can be physically measured in a series of single frames. Additionally, the femto-photographs will show that the two beams of light converge on each other at a relative speed of 2 times C.

Variation on Experiment 1 – Case B

In this variation, the experiment will be configured as to allow for the laser to move along a set of rails towards the stationary target and the oncoming moving target.

When the experiment is run, the frame holding the laser will be set in motion, along with the moving target, and then the laser will fire. What we would expect to find is that the beam of light will be physically shorter then when the laser is stationary. This is because the light originates at a point in space that corresponds to the current location of the exciting mechanism not from the exciting mechanism itself. Because the laser is moving in the same direction, as the light it will be closer to the leading wave front of the beam when the last photon is excited, then it would have been had the laser been stationary.

Variation on Experiment 1 – Case C

This variation is maybe the most interesting, as well as, possibly the most difficult to perform. In this variation of the experiment, we perform the basic experiment and then twist the orientation of the experiment 90 degrees and repeat the experiment. Then twist the experiment another 90 degrees and repeat the experiment, until we have rotated through 360 degrees or compass headings, running the experiment at each heading. Then we will rotate the experiment 90 degrees upward from level, run the experiment, and then rotate the experiment 90 degrees downward from normal and run the experiment.

The purpose of this variation is to demonstrate the movement of the planet through the ocean universe. As the laser is fired along the direction of the movement of the planet through the energy ocean and then against the flow of energy relative to the earth, the speed of light relative to the targets will change accordingly. An alternative variation would be to conduct the experiment at different time of the day while it is in the same orientation to the earth to measure the variations in speed related to the earth's rotation.

The difficulty in this experiment may arise from the earth, like anybody moving through the ocean universe, experiences a bow wave like effect and a microenvironment near its surface as result displacement caused energy density region. As a result of this there may be measured a layer of energy particles moving along with the earth through the ocean universe at the same speed or relatively close speed to that of the earth. Conducting this measurement on the earth's surface may produce a less than satisfactory result, and it may be necessary to conduct this variation from a position of a geosynchronous orbit to get a more precise measurement of the earth's speed through the energy ocean.

Experiment 2 – Measuring Gravity Waves

One of the most exciting activities going on in space exploration right now never leaves the earth. The Search for Extra Terrestrial Intelligence or SETI project is one of the most interesting and provocative undertakings of human history. Yet for all the time, money, and hope that has been invested in this extraordinary undertaking, we have uncovered nothing of alien life, and are continually left with the question of whether all our efforts will ever provide any real value to humanity at all.

Current estimates suggest that the number intelligent civilizations that we might be able to communicate with in our galaxy alone ranges from zero to around a million. If we assume at a number of at least several thousand or tens of thousands, we are still left with the question as to why we have

Fig. 15-2: "Contact", The Search for Extraterrestrial Intelligence, conceived by Carl Sagan (Contact, 1997).

not encountered civilizations more advanced or at least as advanced as we are.

Perhaps the questions we should be asking are really about if we are advanced enough to be in contact with them. If we are to believe the current understanding of Relativity than nothing can travel faster than the speed of light. Meaning, a radio signal sent out from a planet even as close as ten-thousand light years away would take ten-thousand years to reach us. Looking back on our last ten-thousand years of history, we have to question what civilization would have survived long enough to reach one at a stage in technology high enough to signal us, and second, with an expectation of a twenty-thousand year delay in getting a response. The answer is probably very few indeed.

If an advanced civilization on a distant planet did exist and were able to travel through the cosmos how would they get from planet to planet in a time span that would be of any feasible value to them with such large time spans? If they could never dream of visiting us, why would they try to contact us at all? However, what if we are wrong about our perceptions of Relativity? What if the universe is configured differently then we have been led to believe for over a hundred years now? If you could travel through space at speeds in excess of the speed of light, or through wormholes, or other means unknown to us presently, communication between your travelers and home world would be important, but how would that be accomplished? If your space-traveling companions were even a hundred light years from hope, how would you effectively communicate with them using electromagnetic wave communications systems like our own radios?

The answer is actually simple and negates all of our current efforts to communicate with or intercept communications from advanced civilizations using radio (electromagnetic wave) communication systems. Imagine a civilization that had just learned to communicate using mirrors reflecting sun light from mountaintop to mountaintop trying to communicate with a modern aircraft carrier operating beyond their sight horizon. It would not happen, because it would be like the Enterprise on Star Trek using a ham radio to communicate with star fleet rather than subspace communications, it just doesn't work.

In the Star Trek science fiction series, subspace communication (also called subspace radio or the hyperchannel) was the primary form of electromagnetic communication used throughout the Federation. By transmission through subspace rather than normal space, subspace communication permitted the sending of data and messages across interstellar distances faster than the speed of light. This made it much more practical than conventional radio. **In fact, Starships rarely even monitored frequencies that traveled at the speed of light or slower.** (TNG:

"The Ensigns of Command", okudagram; VOY: "The 37's") (Wiki TV, 2015)

Star fleet monitoring communications that are not "subspace" is the equivalent of our SETI program monitoring mirror flashes on mountaintops. To communicate effectively with advance civilizations we need to be able to use a communications technology that is based on a propagation medium that is virtually as instantaneous across the universe as our radio wave communications systems are across town. The purpose of this experiment then is to demonstrate such a capability and open the doors to development of nearly instantaneous communications techniques to communicate across any galactic or intergalactic distance.

Previously we discussed the concept of instantaneousness in regards to the phenomenon of displacement. We also discussed how in the ocean universe displacement of energy particles by particles of mass is fundamental to the concepts of gravity as we know it. If a particle of energy is instantaneously displaced by another particle of energy and it causes yet another particle to be displaced, and this happens to some, then the last particle to be displaced has to have been displaced at the same instant as the first particle. Based on this concept of the ocean universe, we can conclude that an advanced civilization, which was capable of interstellar space travel, would have found a way to harness energy particle displacement waves (gravity waves) for communications, just as we have learned to harness electromagnet waves for our communications purposes.

A simple electro-mechanical device for sensing and magnifying disturbances in the universal gravitational field could be based off the proposed gravitational drive using a rotational body, to create an apparent mass. If we are to spin a gyroscope with an embedded magnet at great speeds, such that the magnet were passed within the field of an electromagnet each revolution, than subtle changes in the magnetic interference between the two magnetics could be created by gravitational wave impacts against the spinning gyroscope, such as the Tajmar effect.

Because of the use of gyroscopic control mechanisms in guidance systems for military missiles and unmanned space craft it has become essential to understand and account for the effects of gravitational anomalies on sophisticated and extremely sensitive gyroscopic systems. Thus, the effects of gravity on gyroscopic systems are among some of the most investigated phenomenon in physics.

In 2007 it was announced by the Space Propulsion group at the Austrian Institute of Technology, that one of the researchers, Martin Tajmar, had identified an anomalous acceleration of a spinning laser gyroscope when it was placed near a rotating super cooled ring. This affect, which has since been labeled the "Tajmar Effect" is very similar to the effect caused on applying an electromagnetic force to a spinning gyroscope as previously discussed in the chapter on propulsion systems (Zyga, 2011). .

More recently in studying the Tajmar Effect, Michael McCulloch, of the University of Plymouth, has verified that the apparent mass of the spinning light beams of the laser gyroscope is affected in some way by being placed next to a spinning super-cooled ring. McCulloch believes that the increase in the rate of spin is a result of the system trying to conserve momentum by increasing speed as apparent mass increases. McColloch feels that this Tajmar Effect may contribute to what has become known as the "Pioneer Anomaly". NASA has recorded that both the Pioneer spacecrafts slowed down more than anticipated as they flew out of the solar system. While McColloch relates this increase in apparent mass to the effects of Unruh radiation, his calculations of the acceleration of the gyroscope at being $\sim 2.67 \times 10\text{-}8$ times the acceleration of the associated super-cooled ring place next to it. This result appears to be consistent with an increase in the energy density surrounding the super-cooled ring as a result of energy particle displacement, consistent with our previous discussion on the topic (McCulloch, 2011).

Of course, in the Ocean Universe Theory we know that the increase in the apparent mass noticed as part of the Tajmar Effect is a direct result of the increase in the density of the localized energy environment around the spinning super cooled ring. As the ring is brought into the area of the laser gyroscope, the light beams/waves are now traveling an increasingly denser medium and like any wave propagation, they will increase in speed. We see the same thing with wave propagations in water, air, and even solids like steel.

McCulloch suggests, *"Inertial mass has not been well understood and has been assumed to be the same as gravitational mass (the Equivalence Principle, EP)... Inertia is important practically since it determines the sensitivity of an object's motion to outside forces."* (McCulloch, 2011). This misunderstanding of how energy exists in the ocean universe is now becoming more problematic to applied science as instrument sensitivity continues to increase, and the energy particle is causing significant variances in operational characteristics. Yet it is these very problematic variations, which are the key to the next generation SETI.

Like Star-Trek, with it "sub-space" communications systems, advanced space traveling civilizations will require a communications system that will allow them to effectively communicate with their spacecraft and distinct colonies. That communications system will rely on gravitational waves created by varying the apparent mass of objects, or in other words, by fluctuating the density of an object's localized energy environment in a pulsating way that will send out gravitational waves, rather than electromagnet waves through the universe. Because gravity waves are a form of displacement, and displacement is instantaneous in nature the gravitational waves are felt across the universe virtually instantaneously. The secret in using them for communications comes in being able not only to create a modulated pattern but also to receive that pattern at the other end. This is where the apparent mass of a mechanical or laser gyroscope comes into play.

By using a device similar to the propulsion devise described earlier, only instrumenting the electromagnetic circuits in a manner as to detect subtitle changes in both the rotational velocity of the ring and the amount of energy being displaced by the spinning ring (apparent mass) one would be able to detect communications signals in the universal gravitational field. By inducing pulsed gravity waves, it would be possible to send digital data across the universe instantaneously for all practical purposes. If a society is advanced enough to travel across the galaxy then they are advance enough to communicate with gravitational waves. They are also advance enough to know not to, or at least not care to, communicate with civilizations that have only reached the technical sophistication of communication with archaic electromagnetic radiation.

Conclusion

In writing this book I set out to accomplish several ends, to fulfil a project that I started in the early 1980s of describing an alternative view of relativity based on the assumption that the E in E=MC^2 held more significance in reality then was being recognized. While not popularly known or regularly taught in our schools, Albert Einstein himself was a firm believer in not only the existence of ether, and in the absolute necessity of a universal ether to maintain the very structure of space-time itself (Einstein, 1920). It seemed that the actual mechanics of not only how light propagates but also how the observable phenomenon of the universe actually came about was continually being lost in the mathematics that seemed to be generated for mathematics sake. I wondered was it possible to start with an observational analysis that utilized a simplified version of formal logic to describe the observable universe. As it turned out, the process resulted in a formation of a single model of the universe that is as applicable at the atomic level as it on the galactic scale.

While we may never know for sure what Einstein ever discovered in his quest for a unified field, the Ocean Universe Theory confirms that his quest was surely not in vain. While work in quantum sciences tends to segregate and compartmentalize phenomena, the Ocean Universe Theory is a natural continuation of relativity that finds its most fundamental assertions in both the concepts of E=MC² and the phenomena of relativity to observers in different frames of motion.

In our exploration of the ocean universe, we have traveled across the cosmos from the depths of the atom here on earth to the ominous galactic centers where the life of the cosmic family of bodies that lights our skies is born. We have travel to the very center of nothingness of the great galactic voids of darkness from where unmeasurable streams of power are shot across the universe. From yesterday to tomorrow, we have followed the transition of spatial information as it propagates across the synaptic creation as structured information packets that record who we were, while defining how our future will unfold. Understanding how time passes has always been on the mind of humanity sense time-immortal but being able to understand what time really is offers a revolutionary understanding of how the history of our universe has unfolded. We can travel to the future or we can revisit our own past, in the same context as accessing data on a memory chip, virtually all things are possible with time. In the ocean universe time becomes a virtual reality that controls and logs all interactions in the universe.

The second accomplishment I set to show with this book and maybe the most important is the very nature of light itself. Once we understand what the speed of light represents then we can see that while it is consistent to itself it is not a finite limit of speed in the universe. For more than a hundred years our science and technology has been limited by a simple mistake in interpreting the data of a single experiment. An experiment that was faithfully repeated numerous times and each time the same data interpretation mistake was made. It has been so entrenched in our education systems that even today, with femto-photography, which can photograph light itself at a trillion frames per second; researchers continue to live with a hundred year old mistake in the design of experiment. Even when todays advanced software, which is intensive and intelligent, offer proofs of the experimental mistake, researchers not only negate the advice of these systems, they actually override the correct analysis and manually input the old mistake.

However, we cannot be too hard on our colleagues of old, as they were in many ways testing theories well beyond the capabilities of their technology to explain. When Albert A. Michelson and Edward W. Morley designed their experiment, they had the benefit of neither electronic communications systems nor advance electronic computers. In 1887 these scientists had only the barest of technology to work with. The speed of sound had been fairly accurately defined almost two-hundred years earlier, when the Reverend William Derham measured the speed of sound at about 1142.5 feet per second within about 15 feet per second of our standard speed of sound of 1,129 feet per second in air at 68 degrees Fahrenheit. In the 1880s Michelson and Morley may not have had the technology they really needed but they still conducted researched to produce science that was experimental based and founded in the principals of natural philosophy.

It was after the turn of the 20th Century that a new twist was added to the realm of modern science, the concept of consensus science. While researcher at the end of the 19th and beginning of 20th Century had to deal with technological limitations that hampered their scientific progress, today's scientific discoveries are being hampered by an even more daunting challenge, "Consensus Science" (Guliuzza, 2009). By design, scientific theory and fact are supposed to be based on empirical evidence, supported by repeated experimentation by different scientists. However, as Dr. Guliuzza suggest, the science community has become a Bastian for the science elitists. Who are these scientific elitist and how did they come to dominate our scientific community? These scientific elitist are individuals who are, well funded, well connected politically, and very charismatic speakers who have been able to funnel significant amounts of limited research funding for their get-along scientific theories. With this new found power they have been able to bully, intimidate, and silence hard experimental science from our universities, public forums, and most dangerously to remove the best proven technologies from our industrial complex where it is costing us both money and technological superiority.

Dr. Guliuzza has this to say about consensus science, *"Scientifically speaking, a serious problem arises when advocates wield "scientific consensus"* as if it were a valid scientific argument that carries the same weight as experimentally-derived evidence-- a practice derisively called 'science by consensuses or 'consensus science.'"* (Guliuzza, 2009). Science without hard and repeatable empirical evidence is nothing more than political debate. Moreover, like politics, consensus science is perfuse with corruption and unjustifiable compromise. William Allman, a science writer, cited by Dr. Guliuzza, says this *"The pressures to publish not only increase the risk of mistakes made in haste but, more menacingly, raise the rewards of outright manipulation of data. Critics argue that the scientific community is generally unprepared to recognize such fraud."* While Nicolas Wade, a researcher on scientific fraud, adds, *"Scientists are trained to believe that research is an entirely objective process....That makes them all the more vulnerable to people who deceive, because they don't have their guard up... As a result, fudged data that conform to prevailing scientific wisdom...can easily slip into print... Advocates of consensus science capitalize on exploiting these problems, not working to fix them."* (Guliuzza, 2009)

For over a hundred years the forces of consensus science have prevented us from evolving scientifically or more accurately transcending into the next higher realm of scientific reality, by continuing to present as verified fact, theories and postulates that are either not supported by independent verification, or are validated with flawed experimental procedures. This unproven science is then popularized for financial gain and fame rather than for true scientific advancement. What does it really matter to humanity?

Today, we stand in awe of the megalithic rockets that have blasted humanity into orbit and beyond, allowing us to walk on Earth's closets neighbor. Yet with approximately 3 millions of pounds of thrust generated by the space shuttle engines it could only carry about 50,000 pounds of payload into orbit. The now antiquated Saturn V rocket generated some 7.6 million pounds of thrust and could lift some 250 thousand pounds of payload into earth orbit. These are impressive numbers for sure. But, what if you could build a craft that could take off like a helicopter and just continue slowly going upward beyond the atmosphere and out beyond the orbit of our satellites and beyond?

Our current growth in technology is bound by the constraints imposed on it by our science, a science that is significantly comprised of individuals regurgitating the work of previous great minds, just as literature critics writing volumes of words of exposé to enhance their position, even above the poet they are writing about. Yet humanity is truly faced with the ultimate flight or fight situation, where the lives of billions of people are at risk. In a finite world we are left with no more physical space to expand and grow as a species here on earth; thus as our flight option seems cut off without the science to escape this world, at some point we will have to initiate our fight reflexes. Understanding the real nature of energy, as a literal an integral part of the universe that takes up space and interacts with physical matter as well as itself, is an important aspect of our universe to understand, if we are to expand the horizons of our science beyond the current constraints it imposes on our technology.

If we can truly understand what is energy is and understand the real power of energy, then we could build crafts that could lift the people and supplies from the earth, carry them to planets both near and far, while being able to communicate with these colonist in a timely and effective way. Getting $E=MC^2$ correct goes much farther then academic honors, it is the means by which humanity is able to transcend and move beyond the current arbitrary boundaries of the space and time it works within. It was not all that long ago that the science of commercial sailing had reached its pinnacle in technological development. Clipper ships like the Cutty Sark were the engineering marvels of their time and were able to achieve great speeds in delivering much need commerce around the world. With vast

The Cutty Sark
(Mason, 1875-1965)

amounts of wealth invested in these graceful works of precession and master craftsmanship, the first clumsy and awkward looking steam powered ships were not a welcome advancement of science and technology to the sailing elitist.

Ultimately, the science of sailing gave way to the science of the steam ship, then to the science of the iron hulled steam ship, until today when the science of the nuclear powered ships for military applications and compressed or liquefied natural gas for commercial shipping has emerged. With the coming of each new generation of science, the existing hierarchy of financial, technical, and operational power structures was overturned. Many leading experts of the time fought against being displaced because of their inability to transcend into the new realms of science that were evolving. Yet once the proverbial "cat was let out of the bag" the inevitable progression of science marched forward, advancing beyond the stagnation of familiarity.

Today the great clipper ships of modern relativistic science are the enormous particle accelerators, the giant radio and optical telescopes, and the powerful scanning tunneling microscopes that see individual atoms. While the next generation science, the steam ships of the future evolution science, is still in its clumsy and awkward adolescent stage, where it is still developing its potential to replace our current science and ascending our technology into a completely new era of functionality. Now that it has been made clear that energy exists as a real and tangible element of the universe, a universe where light is a simple wave action in a medium of energy particles, where traveling faster than the speed of light is no different than breaking the sound barrier, the scientific community cannot stay still or go backwards. Young entrepreneurial scientists will read and comprehend the importance of the concepts presented in this book, translating these concepts into new innovative inventions that will tap resources that will lead to untold wealth. Humanity will not be able to, afford to, or want to stay behind a wall of ignorance now that the massive gate holding them back has been knocked down.

Finally, a third and more recent goal in writing this book is to lay a fundamental foundation in Relativistic Physics, for the concepts of what has over the last several decades has come to be known as the Electric Universe. Whether all the details and constructs of the electric universe theories are ultimately proven, the quest for new scientific knowledge that it has been promoted has the ability to change the future of space physics forever. So what is this concept of an Electric Universe? Its name is given to a collection of scientific findings that indicate that many of the phenomena that we observe in space and contribute to chemical or atomic processes, are in fact the results of electrical processes that can be verified by laboratory and field observations. Experts in this emerging field of cosmology such as David Talbott, Wallace Thornhill, and a host of other internationally recognized academians and researchers have revisited the classical theories for the causes of the tails of comets, the vast gashes and valleys found on mars, and the very cause of radiation of the sun. Each of these phenomena can be equally explained by various forms of electrostatic and plasmatic discharges, as they are by chemical and/or nuclear reactions (Thornhill & Talbott, 2002, 2007). However the realization that energy, or ether as Einstein referred to it, is an active element of our cosmos gives additional credibility in the sense of a physical attribute that can justify the existence of the electrical phenomenon. Can the ocean universe act in a manner similar to an Air-Gap capacitor allowing for large discharges of force between different energy densities of the regions surrounding heavenly bodies.

In this book, we have taken a great journey from the heart of the universe, to the center of a single atom and back again. We have seen how the same forces that hold atoms together are the explanation for the ordering of planets, galaxies, and even the existence of great voids that we often call black holes. We have seen how modern information and digital communications sciences can replace our antiquated analog understanding of time and the question of time traveling. We have explored the proposed speed barrier of light and debunked the flawed experiments that underlie the presumption of such a limit on space travel. We have addressed paradoxes of relativistic speed and proposed solutions to these paradoxes that are firmly grounded in the same physics that describes gravity, light, and mass. The evidence of energy itself, as the ether proposed by Einstein to support the viability of the General and Special theories of Relativity, is as overwhelming evident as the flaws in the design of experiments that have been used to try and measure the relative velocity of light.

Have we answered all the questions? No, obviously not. However, what we have done is laid a foundation for future work in understanding not only the physics of Relativity, but also the very nature of who we are and how we understand at all. The future of research in this area is wide open for those who seek to explore the universe where we have not treaded before. The mathematics to describe the observational analysis needs to be deduced, while, the detailing and implementation of the experiments to verify the theories postulated herein must be accomplished. There are also other aspects of the universe that must be described in a manner consistent with the ocean universe. In this work, we were not able to describe the mechanics of the magnetic fields for example, even though this phenomenon must also conform to the rule of the ocean universe.

What are the next steps? First, to demonstrate through repeatable experimentation that while the speed of light while finite relative to the environment the light is traversing, is not a limiting speed as has been hypothesized for over a hundred years now. A second activity should be experimentation to explore the use of gravitational waves for communications purposes. While a third activity is that of exploring the development of propulsion systems based on the principals of moving through the ambient energy. Will it take time for the acceptance of energy as an actual element of the universe? Of course, many careers and funding for large ongoing experiments are at stake. How do we justify the expenditure of billions of dollars for experiments using the massive accelerators, or the advance satellites, or the years of research into the backing of quantum theory to explain the observable universe at the micro and macro levels with a single scalable explanation? We transition, of course. The failing of a process to produce desired results is not a failing of the people who were tasked to implement it. We need to use the verification of ether, in the context of energy particles as described in The Ocean Universe Theory as rallying point to bring together the great minds of physics. This is especially true of the young minds who have years of professional research ahead of them to dedicate to moving physics and cosmology from the age of "clipper ship" physics/cosmology into the age of "nuclear powered ship" physics/cosmology. This a gigantic step for science but one that offers such a potentially economic advantage to the entrepreneurial elite that the money to fund this research will be forthcoming for those adventurous and visionary physicists who understand the enormous long term benefit to humanity of ascending beyond our current archaic state of scientific understanding.

Is The Ocean Universe Theory valid? It is the ability to ask this question that book is about. While validating this theory could open the door to a new level of scientific thought, the act of testing it through repeatable scientific experimental processes is ultimately the real challenge and test for the scientific community. Do we, as scientists allow the centuries of rigid and proven scientific process to be trodden down by a vogue consensus process? Or, do we hold fast to the proven experimental method? I for one am excited to be experimentally proven right or wrong and throw-down-the-gauntlet to all whom would accept the challenge. If nothing else, I would hope the that the thoughts and analysis portrayed in this book, serve as an inspiration to others to venture out beyond the safety zone of the accepted consensus to ask, "What if?".

Epilog

Albert Einstein

When in the first half of the nineteenth century the far-reaching similarity was revealed which subsists between the properties of light and those of elastic waves in ponderable bodies, the ether hypothesis found fresh support. It appeared beyond question that light must be interpreted as a vibratory process in an elastic, inert medium filling up universal space. It also seemed to be a necessary consequence of the fact that light is capable of polarisation that this medium, the ether, must be of the nature of a solid body, because transverse waves are not possible in a fluid, but only in a solid. Thus *the physicists were bound to arrive at the theory of the ``quas-irigid'' luminiferous ether, the parts of which can carry out no movements relatively to one another except the small movements of deformation which correspond to light-waves...*

This dualism still confronts us in unextenuated form in the theory of Hertz, where matter appears not only as the bearer of velocities, kinetic energy, and mechanical pressures, but also as the bearer of electromagnetic fields. Since such fields also occur in vacuo i.e. in free ether the ether also appears as bearer of electromagnetic fields. The ether appears indistinguishable in its functions from ordinary matter. Within matter it takes part in the motion of matter and in empty space it has everywhere a velocity; so that the ether has a definitely assigned velocity throughout the whole of space. There is no fundamental difference between Hertz's ether and ponderable matter (which in part subsists in the ether)...

The space-time theory and the kinematics of the special theory of relativity were modelled on the Maxwell-Lorentz theory of the electromagnetic field. This theory therefore satisfies the conditions of the special theory of relativity, but when viewed from the latter it acquires a novel aspect. For if K be a system of co-ordinates relatively to which the Lorentzian ether is at rest, the Maxwell-Lorentz equations are valid primarily with reference to K. But by the special theory of relativity the same equations without any change of meaning also hold in relation to any new system of co-ordinates K' which is moving in uniform translation relatively to K. Now comes the anxious question: Why must I in the theory distinguish the K system above all K' systems, which are physically equivalent to it in all respects, by assuming that the ether is at rest relatively to the K system? For the theoretician such an asymmetry in the theoretical structure, with no corresponding asymmetry in the system of experience, is intolerable. If we assume the ether to be at rest relatively to K, but in motion relatively to K', the physical equivalence of K and K' seems to me from the logical standpoint, not indeed downright incorrect, but nevertheless inacceptable.

The next position which it was possible to take up in face of this state of things appeared to be the following. The ether does not exist at all. The electromagnetic fields are not states of a medium, and are not bound down to any bearer, but they are independent realities which are not reducible to anything else, exactly like the atoms of ponderable matter. This conception suggests itself the more readily as, according to Lorentz's theory, electromagnetic radiation, like ponderable matter, brings impulse and energy with it, and as, according to the special theory of relativity, both matter and radiation are but special forms of distributed energy, ponderable mass losing its isolation and appearing as a special form of energy.

More careful reflection teaches us, however, that the special theory of relativity does not compel us to deny ether. We may assume the existence of an ether...

Recapitulating, we may say that according to the general theory of relativity space is endowed with physical qualities; in this sense, therefore, there exists an ether. According to the general theory of relativity space without ether is unthinkable; for in such space there not only wonld be no propagation of light, but also no possibility of existence for standards of space and time (measuring-rods and clocks), nor therefore any space-time intervals in the physical sense. But this ether may not be thought of as endowed with the quality characteristic of ponderable inedia, as consisting of parts which may be tracked through time. The idea of motion may not be applied to it.

- (Einstein, 1920)

Bibliography

Aerospace Web.Org, 2012. *Hypersonic Theory*. [Online] Available at: http://www.aerospaceweb.org/design/waverider/theory .shtml [Accessed 2014].

Aharonov, Y., Cohen, E., Grossman, D. & Elitzu, A. C., 2012. *Can a Future Choice Affect a Past Measurement's Outcome?*. Crete, Physics World.

Alexander, M., 2014. *Swimming Pool Interferometry*. [Online] Available at: http://www.eso.org/public/images/potw1404a/ [Accessed 2015].

Alfvanbeem, 2015. *Here. There. Everywhere.*. [Online] Available at: http://hte.si.edu/own.html [Accessed 2015].

Bell, J. et al., 1999. *Software Design for Reliability and Reuse, A Proof-of-Concept Demonstration,* Portland, OR: Tri Ada '94.

Benson-Avillan, E. M., 2010. *Spectroscopic Analysis of the Dwarf Nova SS Cygni,* Muncie: Ball State University.

Berardelli, P., 2009. *A Flood, Not a Falls, Refilled the Mediterranean.* [Online] Available at: http://news.sciencemag.org/earth/2009/12/flood-not-falls-refilled-mediterranean

Berry, R. S. & Smirnov, B. . M., 2005. Void theory of nucleation in liquids. *Physical Review,* B 72(104201), pp. 1-7.

Bevel, P. J., 2009-2011. *Radiation Pattern.* [Online] Available at: http://www.antenna-theory.com/basics/radPattern.html [Accessed 2015].

BoldMethod.com, 2014. *Vortex Generators: Preventing Stalls At High And Low Speeds.* [Online] Available at: http://www.boldmethod.com/learn-to-fly/aerodynamics/ vortex-generators/ [Accessed 2015].

Bonometti, J. A. & Morton, . P. J., 2000. External Pulsed Plasma Propulsion (EPPP) Analysis Maturation. *American Institute of Aeronautics and Astronautics,* Volume AIAA-2000-3610.

Bose, N., 2014. *Implosions or Explosions,* s.l.: Austrailian Maritime College.

Boyarsky, A., Ruchayskiy, O., Iakubovskyi, D. & Franse, J., 2014. An unidentified line in X-ray spectra of the Andromeda galaxy and Perseus galaxy cluster. *Physics Review Letters,* 15 December.113(25).

Brennen, C. E., 1995. *CHAPTER 3. CAVITATION BUBBLE COLLAPSE.* [Online] Available at: http://authors.library.caltech.edu/25017/4/chap3.htm [Accessed 1 11 2014].

Brown, K. & Wren, C., 2011. *Focus Italy Cavatation Bubble,* s.l.: Mondolithic Studios.

Bulbul, E. 2014. *Perseus Cluster: Mysterious X-ray Signal Intrigues Astronomers.* [Online] Available at: http://chandra.harvard.edu/photo/2014/perseus/ [Accessed 2015].

Buranen, L. & Roy, A. M. eds., 1999. *Perspectives on Plagiarism and Intellectual Property in a Postmodern World.* New York: State University of New York.

Cain, F., 2013. *What Is A Quasar.* [Online] Available at: http://www.universetoday.com/73222/what-is-a-quasar/ [Accessed 13 October 2014].

Cazala, J., 2015. *Architecture-free neural network library.* [Online] Available at: https://www.npmjs.com/package/synaptic [Accessed 2015].

Choi, J., Hsiao, C.-T., Chahine, G. & Ceccio, S., 2009. Growth, oscillation and collapse of vortex cavitation bubbles. *Journal of Fluid Mechanic,* Volume 624, pp. 255-279.

Cline, A., 2012. *Professors on the Take, Selling Out Students to Publishers.* [Online] Available at: http://atheism.about.com/od /aboutethics/a/Professors Bribe.htm [Accessed 2012].

Colebrooke, H. T., 1817. In: *Algebra with Arithmetic of Brahmagupta and Bhaskara..* London: s.n.

Contact. 1997. [Film] Directed by Robert Zemeckis. USA: Warner Bros..

Cosmic Astronomy, 2014. *Bandwidth Definationsof Gravity Waves.* [Online] Available at: http://www.cosmicastronomy.com/blare.htm [Accessed 2015].

Cosmic Astronomy, 2014. *Giant Gravity Waves Pervade Perseus Galaxy Cluster.* [Online] Available at: http://www.cosmicastronomy.com/perseus. htm#perseu29 [Accessed 2015].

CosmosUP, 2014. *The binary system SS Cygni much closer than we thought.* [Online] Available at: http://www.cosmosup.com/the-binary-system-ss-cygni-much-closer-than-we-thought/ [Accessed 2014].

Domain-b.com, 2012. *Gold lenses used to create gamma optics,* s.l.: The Information Company Private Limited.

Dranidis, D. V., 2003. Shipboard Phased Array Radars, Requirements, technology and operational systems. *Waypoint Magazine,* February. Issue 3rd.

Dregely, D. et al., 2011. 3D optical Yagi–Uda nanoantenna array. *Nature Communications,* 05 April.2(Article 267).

Duffy, A., 1999. *Interference of Waves.* [Online]   Available at: http://physics. bu.edu/~duffy/py105/WaveInterference.html [Accessed 1 2014].

Einstein, A., 1920. "Ether and the Theory of Relativity" An Address delivered on May 5th, 1920, in the University of Leyden. In: G. B. Jeffery & W. Perret, eds. *Sidelights on Relativity.* s.l.:Dover Press 1983.

Einstein, A., 1954. Einstein's inaugural address to the Prussian Academy of Sciences, 1914. In: *Ideas and Opinions.* New York: Bonanza Books.

Einstein, A., 1961. *Relativity The Special and The General Theory.* New York, Crown Publishers, Inc..

Einstein, A., n.d. *Relativity The Special and The General Theory.* 1961 ed. New York, NY: Crown Publishers, Inc..

ESA & Bujarrabal, V., 2001. *Hubble reveals previously unseen shocks (heic0111a).* [Online] Available at: http://www.spacetelescope.org/images/heic0111a/ [Accessed 2014].

Fletcher, J. C. & Bailey, R., 1973. *Electromagnetic Wave Energy Converter.* [Online]   Available at: http://patft.uspto.gov/netacgi/nph-Parser?Sect1= PTO1&Sect2=HITOFF&d=ALL&p=1&u=%2Fnetahtml%2FPTO%2Fsrchnum .htm&r=1&f=G&l=50&s1=3760257.PN.&OS=PN/3760257&RS=PN/376025 7 [Accessed 2015].

Florent, G., 2015. *Sailing: From Work To Fun.* [Online] Available at: http://www. planetseed.com/sciencearticle/sailing-work-fun [Accessed 2015].

Free, G. & Gruchy, D., 2011. *Droplet Collisions at 5000fps.* [Online] Available at: https://www.youtube.com/watch?v=cNI-LIVs-to [Accessed 2015 2015].

Frei, W., 2014. *Introducing the Ray Optics Module.* [Online] Available at: https://www.comsol.co.in/blogs/introducing-ray-optics-module/ [Accessed 2015].

Garcia-Castellano, D. et al., 2009. Catastrophic flood of the Mediterranean after the Messinian salinity crisis. *Nature,* 10 December, Volume 462, pp. 778-781.

Gilbert, L., 2012. *Course Theorem.* [Online] Available at: http://www.onemetre.net/Design/CourseTheorem/CourseTheorem.htm [Accessed 2015].

Graepel, T., Herbrich , R. & Botea , A., 2008. *Video Games and Artificial Intelligence.* [Online] Available at: http://research.microsoft.com/en-us/projects/ijcaiigames/ [Accessed 2015].

Guliuzza, R. J., 2009. Consensus Science: The Rise of a Scientific Elite Acts & Facts. *Acts & Facts,* 38(5), p. 4.

Henze, 2009. *Introduction to LIGO & Gravitational Waves.* [Online] Available at: http://www.ligo.org/science/GW-Sources.php [Accessed 2015].

Hutchinson, L., 2013. *How NASA brought the monstrous F-1 "moon rocket" engine back to life.* [Online] Available at: http://arstechnica.com/science /2013/04/how-nasa-brought-the-monstrous-f-1-moon-rocket-back-to-life/ [Accessed 2015].

Janssen, P., 2009. Wave Interactions. In: *The Interaction of Ocean Waves and Wind.* West Nyack, NY: Cambridge University Press, pp. 209-274.

jinnwe.com, 2014. *process of creating a water jet from an imploding cavitation bubble,* s.l.: s.n.

Kaushal, R. S., 1998. Existence of a space invariant in the Tolman-Oppenheimer-Volkoff theory. *Classical and Quantum Gravity,* 15(1), p. 197.

Kraus, J. D., 1956. *Spherical cage antenna.* [Online] Available at: http://www.google.com/patents/US2732551 [Accessed 2015].

Kurz, T., Metten, B., Schanz, D. & Lauterborn, W., 2008. *MOLECULAR DYNAMICS SIMULATIONS OF BUBBLE COLLAPSE.* [Online] Available at: http://webistem.com/acoustics2008/acoustics2008/cd1/data/fa2002-sevilla/forumacusticum/archivos/non01001.pdf [Accessed 2014].

Lauterborn, W. & Bolle, H., 1975. Experimental investigations of cavitation-bubble collapse in the neighbourhood of a solid boundary. *Journal of Fluid Mechanics,* 72(part 2), pp. 391-399.

Liebeherr, J., Burchard, A. & Ciucu, F., 2010. *Non-asymptotic Delay Bounds for Networks with Heavy-Tailed Traffic.* s.l., IEEE Communications Society.

Lin, G. H., Reyimjan, A. & Bockris, J. . O., 1996. Investigation of resonance light absorption and rectification by subnanostructures. *Journal of Applied Physics,* 1 July, 80(1), p. 565–568.

Luard, L., 2014. *Sea Anchor Basics And Why You Should Carry One.* [Online] Available at: http://www.sea-anchors.com/sportfishing.html [Accessed 2015].

Lu, X., 2005. *Universe, Phase Transitions in Early,* Champaign, IL: University of Illinois Urbana-Champaign.

Lynch, J., 2006. *http://www.writing-world.com/rights/lynch.shtml.* [Online]
Available at: http://www.writing-world.com/rights/lynch.shtml [Accessed
12 June 2012].

MacGregor, M. H., 1992. In: *The Enigmatic Electron.* Boston: Klurer Academic,
pp. 4-5.

Maksymov, I. S., Miroshnichenko, A. E. & Kivshar, Y. S., 2012. Actively tunable
optical Yagi_Uda nanoantenna with bistable emission characteristics. *Optics
Express,* February, 20(8), pp. 8929-8938.

Marks, A. M., 1984. *Device for conversion of light power to electric power.*
[Online]   Available at: http://patft.uspto.gov/netacgi/nph-Parser?Sect1=
PTO1&Sect2=HITOFF&d=PALL&p=1&u=%2Fnetahtml%2FPTO%2Fsrchnu
m.htm&r=1&f=G&l=50&s1=4445050.PN.&OS=PN/4445050&RS=PN/44450
50 [Accessed 2015].

Mason, F. H., 1875-1965. *Cutty Sark.* [Art] (Royal Academy).

McCulloch, M. E., 2011. The Tajmar effect from quantised inertia.. *EPL
(Europhysics Letters),* August, 95(3), p. 39002.

Michelson, A. A., 1887. On the Relative Motion of the Earth and the Luminiferous
Ether. November, CXXXIV(203).

Miller-Jones, J. C. A. et al., 2013. An Accurate Geometric Distance to the Compact
Binary SS Cygni Vindicates Accretion Disc Theory. http://www.sciencemag
.org/content/340/6135/950.full, 24 May, 340(6135), pp. 950-952.

Milonni, P., 1994. *The Quantum Vacuum. An Introduction to Quantum
Electrodynamics.* Boston(MA): Academic Press, Inc.

Mirecki, A., 2011. *Solar sail.* [Online] Available at: http://en.wikipedia
.org/wiki/Solar_sail [Accessed 2015].

Moroianu, A. & Moroianu , S., 2010. *Ricci Surfaces,* Pittsburgh, PA: citeseerx.

NASA, 2012. *Dark Energy, Dark Matter,* Washington, DC: National Aeronautics
and Space Administration.

Nave, C. R., 2005. *Bulk Elastic Properties.* [Online] Available at: http://
hyperphysics.phy-astr.gsu.edu/hbase/permot3.html [Accessed 2015].

Nave,   R.,   2000.   *Precession   Torque.*   [Online]   Available   at:
http://hyperphysics.phy-astr.gsu.edu/hbase/rotv2.html#rvec6

Newton, I., 1729-1968. System of the World, Rules Of Reasoning In Philosophy.
In: *Mathematical Principles of Natural Philosophy.* London: Dawsons of Pall
Mall.

Observatori Ventalló, 2009. *NGC 6960, Western Veil Nebula.* [Online] Available
at: http://observatoriventallo.wordpress.com/2009/06/24/ngc-6960-
western-veil-nebula/ [Accessed 2014].

Overduin,   J.,   2007.   *Einstein's   Spacetime.*   [Online]   Available   at:
http://einstein.stanford.edu/SPACETIME/spacetime2.html   [Accessed   20
August 2012].

PASCO,   1996-2015.   *Speed   of   Light   Experiment.*   [Online]   Available   at:
http://www.pasco.com/prodCatalog/EX/EX-9932_speed-of-light-
experiment/ [Accessed 2015].

Pauling, L., 1964. In: *College Chemistry.* s.l.:Freeman, p. 57.

Pennicott, K., 2002. *Bubble bursts for 'sono-fusion',* s.l.: Physics World.

Physics.Org, 2005. *Temperature inside collapsing bubble four times that of sun.* [Online] Available at: http://phys.org/news3229.html#nRlv

Rambelya, F. d. l., 2010. *Hydrogen Emission and Absorption Spectrum.* [Online] Available at: http://socphysics.blogspot.com/2010/08/hydrogen-emission-and-absorption.html [Accessed 2015].

Raskar, R., 2012. *Imaging at a trillion frames per second.* [Online] Available at: https://www.youtube.com/watch?v=Y_9vd4HWlVA [Accessed 2015].

RedOrbit, 2009. *Flooding Caused Mediterranean Sea To Fill In 2 Years.* [Online] Available at: http://www.redorbit.com/news/science/1797789/flooding_caused_mediterranean_sea_to_fill_in_2_years/

Riemer-Sorensen, S., 2014. Questioning a 3.5 keV dark matter emission line. *arXiv.org,* 5 June.

Rojas, G. d., 1984. *Chapter 3, The Trivalent Logic of Aymara.* [Online] Available at: http://www.aymara.org/biblio/html/igr/igr3.html [Accessed 01 02 2015].

Rose, N., 2011. *SF Fleet Week 2011 13.* [Online] Available at: https://www.flickr.com/photos/31816607@N04/6227398869 [Accessed 2014].

Schoolmasters' Science, 2014. *Ptolemaic System Model.* [Online] Available at: http://www.schoolmasters.com/categories/productDetails.cfm?product_ID=07749&div=sc

Schulze, J. & Sesterhenn, J., 2007. *iSGTW Image of the week - Acoustic field of a supersonic jet.* [Online] Available at: http://www.isgtw.org/visualization/isgtw-image-week-acoustic-field-supersonic-jet [Accessed 2014].

Schwarzschild, K., 2008. On the Gravitational Field of a Point-Mass, According to Einstein's Theory. *The Abraham Zelmanov Journal, The journal for General Relativity, gravitation and cosmology,* Volume 1, pp. 10-19.

Schwarzschild, K., 2008. On the Permissible Numerical Value of the Curvature of Space. *The Abraham Zelmanov Journal, The journal for General Relativity, gravitation and cosmology,* Volume 1, pp. 64-73.

Science TV, 2010. *Time Dilation | Einstein's Relativity.* [Online] Available at: https://www.youtube.com/watch?v=G-R8LGy-OVs [Accessed 2015].

Scoles, S., Calçada, L. & Napier, B., 2012. *Fast Stats (from Space).* [Online] Available at: http://smallerquestions.squarespace.com/blog/2012 /11/30/fast-stats-from-space.html [Accessed 2014].

Scott, J., 2005. *Golf Ball Dimples & Drag.* [Online] Available at: http://www.aerospaceweb.org/question/aerodynamics/q0215.shtml [Accessed 2015].

Seboldt, W. & Dachwald, B., 2002. *SOLAR SAILS FOR NEAR-TERM ADVANCED SCIENTIFIC DEEP SPACE MISSIONS.* Milano, Politechnico di Milano, p. 49.

Shanker, V. K., 2012. *Seeking the center.* [Online] Available at: http://www. eecis.udel.edu/~vijay/BLAST/surface_tension /cork.html [Accessed 2015].

Shooter, S. D., 2015. *Last surfer girl of this summer?.* [Art].

Skrutskie, M., 2005. *Light Absorption.* [Online] Available at: http://www.astro. virginia.edu/class/ skrutskie/images/ [Accessed 2015].

Sonja, 2014. *In Defense of Ptolemy.* [Online] Available at: http://www.salonsonja.com/itinerary-later-on.html

Stannered, 2012. *Double Interference Experiment schematics with correct interference pattern.* [Online] Available at: http://en.wikipedia.org/ wiki/Young%27s_interference_experiment#/media/File:Ebohr1_IP.svg [Accessed 2015].

Steinhardt, P. J., 1981. Monopole dissociation in the early Universe. *Physical Review D. particles, fields, gravitation, and cosmology,* 15 August, Volume 24, p. 842.

Stolfi, J., 2011. *Strong Interaction.* [Online] Available at: http://en.wikipedia.org/wiki/Strong_interaction [Accessed 2015].

Strong, S., 2014. *Tales of the Sea: The Untold History of Aboriginal Ocean-Going.* [Online] Available at: http://wakeup-world.com/2014/03/20/tales-of-the-sea-the-untold-history-of-aboriginal-ocean-going/ [Accessed 2015].

Tann, 2014. Ancient era of fast growth in supermassive black holes studied. *News Network Archeology.*

The McGraw-Hill Companies, 2002. A15 Phases. In: *McGraw-Hill Concise Encyclopedia of Physics.* s.l.:The McGraw-Hill Companies, Inc.

Thompson, C., 2006. *Shock waves made visible..* [Online] Available at: http://www.collisiondetection.net/mt/archives/2006/01/_shock_waves_fr .php

Thornhill, W. & Talbott, D., 2002, 2007. *The Electric Universe.* Portland, Or: Mikamar Publishing.

UCLA Physics Department, 2014. *ePhysics.* [Online] Available at: http://www.physics.ucla.edu/demoweb/newtables/energy _scales.htm [Accessed 2014].

University of Cambridge, 2010. *Mysterious Movements - Surface Tension and a Ball.* [Online] Available at: http://www.thenakedscientists.com/HTML /experiments/exp/mysterious-movements-surface-tension-and-a-ball/ [Accessed 2015].

Valente, M. B., 2010. *The concept of vacuum in quantum electrodynamics,* Pittsburgh, PA: PhilSci Archieve, University of Pittsburgh.

Veldes, G. P. et al., 2013. Electromagnetic rogue waves in beam–plasma interactions. *Journal of Optics,* 4 June .15(6).

Vlachos, T., 1999. Minimal Surfaces in a Sphere and the Ricci Condition. *Annals of Global Analysis and Geometry,* April, 17(2), pp. 129-150.

Wiki TV, 2015. *Subspace communication.* [Online] Available at: http://en. memory-alpha.org/wiki/Subspace_communication [Accessed 2015].

Wikimedia, 2014. *Wave Diffraction.* [Online]
Available at: http://upload.wikimedia.org/wikipedia/commons/3/3c /Wave_Diffraction_4Lambda_Slit.png [Accessed 2015].

Wikipedia, 2015. *Gyro Monorail.* [Online] Available at: http://en.wikipedia.org/ wiki/Gyro_monorail

Wikipedia, 2015. *Sailing Faster than the Wind.* [Online] Available at: http://en. wikipedia.org/wiki/Sailing_faster_than_the_wind [Accessed 2015].

Will, F., 1963. Publica Materies. *Arion*, 2, No. 4(Winter 1963), pp. 131-142.

Williams, M., 2011. *on January 18, 2011.* [Online] Available at: http://www. universetoday.com/83380/double-slit-experiment/ Accessed 2015].

Young., T., 1807. *A course of lectures on natural philosophy and the mechanical arts..* Bedford Bury: William Savage.

Zyga, L., 2011. Gyroscope's unexplained acceleration may be due to modified inertia. *Physics.Org,* 26 July.

www.ingramcontent.com/pod-product-compliance
Lightning Source LLC
Chambersburg PA
CBHW051852170526
45168CB00001B/81